APELL for Mining

Guidance for the Mining Industry in Raising Awareness and Preparedness for Emergencies at Local Level

Orchard swamped by tailings, Aznalcóllar, Spain

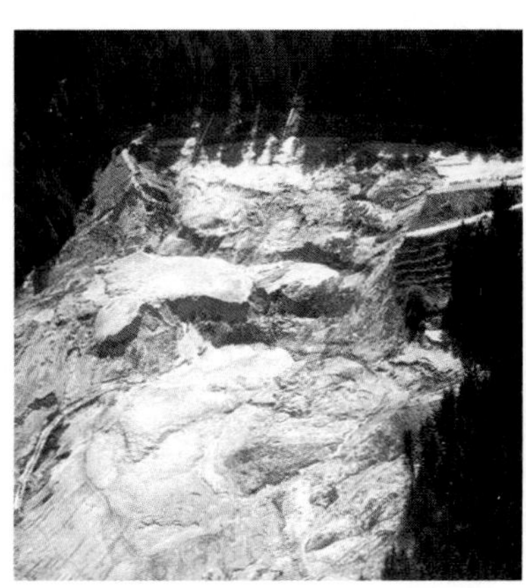
Fatal failure of tailings dams at Stava, Italy

Fish killed by cyanide spill at Baia Mare, Romania

ICME

Printed with the assistance of the
International Council on
Metals and the Environment

UNEP

Division of Technology, Industry and Economics
Production and Consumption
Tour Mirabeau
39/43 quai André Citroën
75739 Paris Cedex 15
FRANCE
Tel: +33 1 44 37 14 50
FAX: +33 1 44 37 14 74
E-mail: unep.tie@unep.fr
www.uneptie.org

Contents

Quotes in support

> *'We are living in a period in which the various social partners are re-evaluating their contribution to sustainable development. It is heartening to see industry at the forefront of this trend, nowhere more so than in the area of environmental emergencies. All affected parties need to be drawn into a common policy of safeguarding our communities and our environment from the risks which modern society inevitably presents. A consultative, preventive approach can be the only rational basis for building a sustainable future.'*
>
> Klaus Töpfer, UNEP Executive Director

'At Rio Tinto we aim to prevent any accident that may cause harm to people or the environment. We also require that our operations, in cooperation with their local communities, prepare, test and maintain procedures that will deal with any emergency situation should it arise. UNEP's APELL for Mining Handbook will greatly assist us, and other mining companies, in engaging more effectively with local people in our emergency planning.'

Leigh Clifford
Chief Executive, Rio Tinto

'As an industry, our responsibility for emergency preparedness and response capacity does not stop at the mine gate. It is a truism of modern mining that we don't operate in isolation, especially when it comes to the safety and well-being of our surrounding communities. This APELL for Mining Handbook provides an indispensable resource for strengthening the industry's practices in this important area.'

John Carrington
Vice-Chairman and Chief Operating Officer,
Barrick Gold Corporation

'When the chemical industry and UNEP together developed the APELL approach in 1988, we had little idea of how far it would go. In addition to APELL for fixed installations, we have seen it applied to transport, to port areas and now to mining operations. I am delighted to recommend this Handbook to the mining industry to help ensure that community protection and environmental protection go hand in hand with economic performance.'

Stanley Szymanski
Responsible Care Leadership Group Chair,
International Council of Chemical Associations

'The issue of emergency preparedness is a crucial one for the mining industry. The International Council on Metals and the Environment strongly endorses the approach taken by UNEP in APELL for Mining—it will definitely assist the industry in mitigating the number and impacts of emergency events.'

Gary Nash
Secretary General,
International Council on Metals and the Environment

Acknowledgements

The United Nations Environment Programme (UNEP) has prepared this *APELL for Mining* Handbook to contribute to improved emergency preparedness in mining companies and mining communities. The publication uses expertise developed by UNEP in cooperation with the chemical industry.

There has been considerable interest in the initiative. At a workshop held in May 2000 by UNEP, in conjunction with the International Council on Metals and the Environment (ICME), the participants agreed on the importance of improving emergency preparedness in the mining industry, particularly in relation to potentially affected communities. ICME has welcomed the development of this Handbook and will work with member companies to encourage its use. Interest is being received from mine managers keen to use it at their operations. Some governments have also expressed interest in using the Handbook to help carry out emergency preparedness programmes in regions with identified high-risk sites.

Once it is tested and tried in the field, the Handbook should be reviewed and, if necessary, refined to maximize its practical utility for the mining industry and adjacent communities.

Many individuals and companies assisted in the development of this Handbook, and UNEP would like to thank them for their involvement. Kathryn Tayles, Senior Industry Consultant, made available through the Minerals Council of Australia, was responsible for the compilation of the Handbook. The first drafts were prepared by Kate Harcourt, Mining Environmental Consultant. The following people acted as reviewers of the document and provided invaluable comments and suggestions:

- Carlos Aranda, Chairman, Environmental Committee, National Society of Mining, Petroleum and Energy, Peru
- Steven Botts, Vice President, Environmental Health & Safety, Compañia Minera Antamina
- Christine Burnup, CEO, Australian Minerals and Energy Environment Foundation
- Alan Emery, Head of Health, Safety & Environment, Rio Tinto
- Timothy Gablehouse, Colorado Emergency Planning, USA
- K.C. Gupta, Director General, National Safety Council, India

- Xia Kunbao, Coordinator, Emergency Response, UNEP
- Lars-Åke Lindahl, Vice President, Environmental Affairs, Boliden
- Jim Makris, Director CEPPA (Chemical Emergency Preparedness & Prevention), Office of Solid Waste & Emergency Response, US EPA
- John McDonough, Vice President, Environment, Barrick Gold Corporation
- James Cooney, General Manager, Strategic Issues, Placer Dome
- Peter Nicholl, HSE Executive, Billiton Base Metals
- Ed Routledge, HSE Manager, Billiton

UNEP gratefully acknowledges contributions of case studies from the following companies:
- Aurul
- Barrick Gold Corporation
- Boliden
- Minera Yanacocha
- Placer Dome
- Rio Tinto

Within the UNEP Division of Technology, Industry and Economics, the following people contributed to the Handbook:
- Fritz Balkau, Chief, Production and Consumption
- Dr Ernst Goldschmitt, Senior Industry Consultant, APELL Programme
- Wanda Hoskin, Senior Programme Officer, Mining
- Barbara Huber, Secretary, Production and Consumption

- Geoffrey Bird, Editorial Consultant

Preface: why read this book?

Recent accidents at mines have had extensive public press. In many cases the response bodies, the community at large, and even the companies, were not fully prepared to deal with such emergencies. In the information vacuum that ensued, inappropriate communication and action (or possibly non-action) took place.

While this media exposure partly reflects a public fascination with disasters and a certain antipathy towards mining, we should all ask ourselves whether there is fire behind the smoke. UNEP believes there is, and that more visible responsibility on the part of the industry for the impact of its operations is a prerequisite for building public confidence. This is especially important at a time when the industry is taking steps to improve its role as a responsible community partner for sustainable development.

To move forward, a number of issues have to be faced.

Safety performance in the industry is uneven. It varies among companies and countries. Official statistics are only the tip of the iceberg. We know that many stories of accidents, near-misses, and 'accidents waiting to happen' go untold. If we understate the magnitude of the real risks, our programmes will miss their target. Despite the many positive developments in the industry, lack of communication on safety aspects and their public dimension is an uncomfortable legacy of past attitudes.

We know that many risk situations extend beyond the boundaries of a site. Towns, villages, rivers, wetlands, farmlands and roads are all potential 'risk objects', to use a term from the emergency response profession. Some risks can no longer be externalized. Nearly all mines have extended transport links for fuel, chemicals, products and perhaps even wastes. The world at large now expects these risks to have an owner, and companies are also increasingly being held responsible for the performance of their contractors. Many external impacts have not in the past been included in Health, Safety and Environment programmes—they must now be more generally included.

These aspects are not limited to operating mines. The industry is generating a growing number of closed operations, rehabilitated tailings and waste dumps, pits and voids. These comprise an ever-expanding inventory of sites which could generate accidents in the future when the companies have departed and only the communities remain. Data shows that the majority of recorded accidents occur at operating sites, but there is no room for complacency. The potential for post-closure accidents needs more consideration especially when the permanent nature of these sites is taken into account. Public awareness of and preparedness for such possible hazards must be an element in future mining policy.

Many mining companies have taken serious steps in recent years to meet the challenge of the environmental and community agendas. Increasingly, major companies are addressing the emergency response issue. The primary focus of companies has been on, on-site action, on risk reduction and emergency response. This is no longer enough. The public aspect is now important. This publication presents an approach for addressing that dimension based on the proven logic of UNEP's APELL programme.

Jacqueline Aloisi de Larderel
Director, Division of Technology, Industry and Economics, UNEP

May 2001

Introduction: scope and application of this handbook

This *APELL for Mining* Handbook provides a framework for the preparation of an Emergency Response Plan that can be used by mine management, emergency response agencies, government officials and local communities. It introduces the generic objectives and organizational framework of UNEP's Awareness and Preparedness for Emergencies at Local Level (APELL) programme, covers risk factors specific to the mining industry, and describes how APELL can be applied to the mining industry. Case studies illustrating the consequences of mining accidents are also presented, underscoring the vital nature of sound preparedness for emergencies.

The Handbook will be of assistance to mining companies and communities in two ways. Firstly, against the backdrop of accidents and risks in the mining industry, it will help to raise awareness of the importance of preparedness for emergencies at local level, within both companies and communities. Secondly, it will help companies, communities and emergency response providers to become thoroughly prepared for the work required.

Within the mining industry, a number of recent tailings spills and other accidents have caused serious environmental and community impacts. Given the number of mining operations around the world, major accidents are relatively infrequent, but they are continuing to occur with unacceptable regularity. Over the last 25 years there have been some 33 major accidents world-wide resulting in releases to the environment. Of these, seven have resulted in fatalities in adjacent communities while others have caused physical damage to property and farmland. Where chemicals have been released, fish and other species have been killed and human health and livelihoods threatened. Accidents have had serious financial consequences for companies and have also seriously damaged the image of the industry as a whole.

Thus mining, like many other industries, faces the challenge of doing more to prevent accidents and to ensure that contingency planning, awareness and communication reduces their impact. Communities and governments expect this. It is an important element in the industry's accountability to the communities within which it operates, and a key determinant of its reputation and acceptability. *APELL for Mining* will help companies to meet that expectation.

APELL concepts are equally relevant to mines and to refineries and smelters, but this Handbook concentrates on mining situations. Downstream mineral processing facilities can pose risks, and accidents do occur, but emergency plans tend to be in place and the impacts of accidents at refineries or smelters have largely been contained on site. Refineries and smelters are essentially large industrial plants for which the original basic APELL programme remains relevant. This Handbook therefore uses examples or lists factors most relevant to mining operations and accidents. These may result from human factors or from natural events. For example, seismic activity or intense rainfall can destabilize waste dumps and tailings dams or cause subsidence; a forest fire can cause fuel tanks to explode. The Handbook also focuses on tailings dam spills, because of their frequency and the potential severity of their consequences.

There have been significant transport accidents involving spills of chemicals to or from mine sites, and these also raise challenges, since by definition there is no site boundary in these cases. This complicates the building of relationships, communications and emergency plans along the transport chain. *TransAPELL*, which extends the APELL concept to risks arising from the transport of dangerous goods, provides more specific guidance in this area. *APELL for Port Areas* is also relevant for the mining industry in some instances.

For communities adjacent to mining operations, *APELL for Mining* will help them better understand those operations and the risks they entail. Experience has shown that local communities are often inadequately informed of risks in their midst and unprepared for emergencies. A fast and effective local response to an accident can be the most important factor in limiting injury to people and damage to property and the environment. While accidents destroy community confidence, a well-informed, well-prepared community is better able to deal with their aftermath.

Given the wide differences in community infrastructure, response capabilities, risks, scale, resources and regulatory frameworks which exist, application of APELL will be unique to each operation and each community. In many places the Handbook gives lists of ideas or suggestions for consideration. Operations should select those elements which are relevant to their situation, and use them as an aid rather than a prescription.

And finally, focussing on pragmatic preparation for emergencies in the event that they may happen should not detract from the pre-eminent goal of accident prevention. As United Nations Secretary General Kofi Annan said in July 1999,

'Prevention is not only more humane than cure; it is also much cheaper. Above all let us not forget that disaster prevention is a moral imperative, no less than reducing the risks of war'.

Disasters cannot be totally eliminated and preparing for them and reducing their impacts is also a moral imperative.

Who should use the Handbook?

- Consistent with UNEP's aim to encourage companies to be proactive in emergency planning, **mine management** is seen as the primary client for this *APELL for Mining* Handbook. It aims to assist operational managers as well as corporate staff responsible for emergency planning to approach the task without having to think through every aspect from scratch.
- **Emergency response agencies and community groups** may initiate the APELL process, perhaps as a result of learning about an

accident at a mine elsewhere. The Handbook can also be used by these organizations.

- **Governments** can use the Handbook as an educative tool to raise awareness and catalyze reviews of plans and programmes in the minerals sector or in areas with a concentration of high-risk operations. They can use also it to ensure that they are prepared to act effectively in the case of an emergency and that their plans are coordinated with others and well communicated.
- **Industry associations** can use the Handbook to highlight the importance of emergency preparedness with their members, and to provide assistance to help companies tackle the work.
- **Within companies**, external affairs, community relations, communications and environmental, health and safety specialists also need to be prepared for emergencies, and this Handbook should be of assistance to them. It gives some advice on the communications aspects which are integral to effective emergency response. It does not duplicate the detailed guidance in the areas of risk communication and media relations which is available elsewhere.

APELL and transboundary situations

The implementation of emergency preparedness planning may involve people and communities across local, regional and international borders. On the one hand this will create complexities in undertaking the process. On the other, the fact that different jurisdictions are involved with different emergency systems, resources, communications and sometimes different languages, reinforces the need for an effective emergency preparedness process in order to formulate a coordinated response plan which can work regardless of boundaries.

APELL and small scale operations

Large and small mining operations should be equally concerned with contingency planning and with being fully prepared for emergencies. The APELL process can be applied at a scale and manner commensurate with the risks involved and resources available. Nevertheless, it is likely that there is a certain size of mine below which a formal APELL process would be difficult to implement. Small scale and artisanal mining operations have the potential to cause off-site damage, particularly when they are concentrated in an area, but the organization and resources required to move through the formal APELL process may be lacking. In some areas, the activities of artisanal miners are becoming centralized through cooperatives, in which case introducing the APELL process may be feasible. Governments or industry associations may need to take the lead with outreach programmes and assistance to small miners regarding emergency preparedness—individually, or collectively in a geographic region.

Section 1

What is APELL? • Objectives • Introduction to the APELL Process and Co-ordinating Group • Responsibilities • Benefits to mining companies and communities

What is APELL?

The Awareness and Preparedness for Emergencies at Local Level (APELL) programme is a process which helps people prevent, prepare for and respond appropriately to accidents and emergencies. APELL was developed by the United Nations Environment Programme, in partnership with industry associations, communities and governments following some major industrial accidents that had serious impacts on health and the environment. APELL is now being implemented in nearly 30 countries around the world.

The APELL Handbook, launched in 1988, sets out a ten-step process for the development of an integrated and functional emergency response plan involving local communities, governments, emergency responders and others. This process creates awareness of hazards in communities close to industrial facilities, encourages risk reduction and mitigation, and develops preparedness for emergency response.

APELL was originally developed to cover risks arising from fixed installations, but it has also been adapted for specific applications: *APELL for Port Areas* was released in 1996, and *TransAPELL, Guidance for Dangerous Goods Transport: Emergency Planning in a Local Community* was published in 2000.

Basically, APELL is a tool for bringing people together to allow effective communication about risks and emergency response. The process of dialogue should help to:

- reduce risk;
- improve effectiveness of response to accidents;
- allow ordinary people to react appropriately during emergencies.

In the case of mining, communication is between the three main groups of stakeholders—company, community, and local authorities. Discussion on hazards usually leads to the identification of risk reduction measures, thus making the area safer than before. Structured communication between emergency response bodies (public and company) results in a better-organized overall emergency response effort.

None of the elements of APELL is radical or new. The programme simply provides a common-sense approach to accident prevention and response. APELL can apply to any risk situation, whether industrial or natural. It can be initiated by any party, although companies can be expected to take the lead. It can be facilitated by governments, or by industry associations. APELL can be applied in developed and developing countries and in remote or urban areas.

What are the objectives of APELL?

APELL's overall goal is to prevent loss of life or damage to health and social well-being, avoid property damage, and ensure environmental safety in a local community.

Its specific objectives are to:

- provide information to the concerned members of a community on the hazards involved in industrial operations in its neighbourhood, and on the measures taken to reduce risks;
- review, update, or establish emergency response plans in the local area;
- increase local industry involvement in community awareness and emergency response planning;
- integrate industry emergency plans and local emergency response plans into one overall plan for the community to handle all types of emergencies; and
- involve members of the local community in the development, testing and implementation of the overall emergency response plan.

The APELL Process

The 10 steps to APELL as given in the *'Handbook on Awareness and Preparedness for Emergencies at Local Level'*, the original APELL document, are shown opposite.

These steps are further developed in Section 4 of this Handbook.

The APELL Co-ordinating Group

Establishing a formal Co-ordinating Group is a key part of the implementation of the APELL process. The Co-ordinating Group provides a mechanism for interaction and cooperation between the many players involved in preventing or responding to emergencies—management, local authorities, community leaders. It provides a means to achieve a coordinated approach to emergency response planning and to communications within the community. It can gather facts and opinions, assess risks, establish priorities, identify resources for emergency response, evaluate approaches, and enhance communication. It can draw in the right people and resources to make sure that following the APELL steps will produce good results.

The Co-ordinating Group is discussed in more detail with special reference to mining in Section 3 of this Handbook.

The ten steps of APELL

Step 1	Identify the emergency response participants and establish their roles, resources and concerns.
Step 2	Evaluate the risks and hazards that may result in emergency situations in the community and define options for risk reduction.
Step 3	Have participants review their own emergency plan for adequacy relative to a coordinated response, including the adequacy of communication plans.
Step 4	Identify the required response tasks not covered by the existing plans.
Step 5	Match these tasks to the resources available from the identified participants.
Step 6	Make the changes necessary to improve existing plans, integrate them into an overall emergency response and communication plan and gain agreement.
Step 7	Commit the integrated plan to writing and obtain approvals from local governments.
Step 8	Communicate the integrated plan to participating groups and ensure that all emergency responders are trained.
Step 9	Establish procedures for periodic testing, review and updating of the plan.
Step 10	Communicate the integrated plan to the general community.

Responsibility for emergency preparedness and response

All parties share an interest in preventing accidents and in minimizing their damage.

- **Companies** have a responsibility both to minimize risks and to ensure effective planning for response, even though it would normally be government agencies which have the statutory accountability for emergencies both outside and inside the boundaries of industrial facilities.
- **National governments** and local agencies will have different, complementary accountabilities. It is a responsibility of government to be prepared and act effectively in the event of public emergencies.
- **Communities** share a responsibility to be prepared and to take action on their own behalf. Community leaders and community organizations should take an interest in the hazards to which they are exposed and the protective measures to which they can contribute.

Leading companies are extending their emergency response planning to include scenario plans based on significant potential environmental incidents. This is increasingly expected and some jurisdictions now require it by law. Also, in many places where the mining industry operates, it is the company which has the skills, equipment, management and communication capacity to make the difference in reducing accidents or their impacts.

Outcomes are the important thing and this Handbook encourages companies in particular to be proactive in working with communities and government agencies to ensure that good emergency planning is in place.

Any operation that has facilities which pose significant risks to nearby communities or sensitive environments should plan for emergencies. The extent of the damage resulting from accidents depends partly on the nature of the immediate response to the emergency at the scene of the accident and in the adjacent community. People likely to be affected by an accident need to know what they, as individuals, should do.

An inappropriate response to an incident because of lack of knowledge or an incomplete understanding of the risks can transform a minor incident into a crisis. It is essential to provide communities with information relating to hazards they may face before an incident occurs. While industry is often uncomfortable with raising the spectre that things could possibly go wrong, this is nevertheless the best way to reduce the likelihood that communities will be placed at risk through their incorrect actions when faced with an incident, or to reduce the likelihood that they will over-react.

> *'Crisis management also contains an ethical component. Although organizations may not be legally liable for certain calamities, that does not remove the moral imperative to identify potential threats and take preventative actions.'*
>
> David W Guth, Proactive Crisis Communication, Communication World, June/July 1995

The benefits of implementing APELL

Most mining companies have on-site emergency response plans which they test from time to time. Preparedness to deal with emergencies involving off-site impacts is usually not as advanced although such accidents have the potential to be more significant in terms of damage and reputation and are often more complex to deal with.

APELL can be useful in any situation that requires joint planning by several parties to develop integrated and well understood response plans ready to be implemented should an accident occur.

The APELL process should bring benefits in at least three ways:

- In **reducing the likelihood** of accidents and **reducing their impacts**. Even if risks are believed to be low, the consequences to a company of a major accident can be severe in physical, financial and reputational terms. APELL can help protect the company as well as the community.
- In helping to **build relationships** between the mine and the community which will be of benefit over the long term. Mining companies are becoming more transparent, proactive and responsive in their relationships with stakeholders. Emergency preparedness planning requires effective communication between all parties, which helps to build relationships based on common interest.
- In assisting community awareness and understanding of the operation and its management which should generate the **confidence, trust and support** which companies need whether or not they experience an accident. These will be severely tested if there is a major accident, but if trust exists, the company will be better placed to communicate effectively in the case of an emergency as well as to recover more quickly from one.

> *'Co-operation between stakeholders should be developed before an incident, not after. That would give a chance to establish roles and responsibilities, action plans, etc. and also to build trust between parties.'*
>
> Lars-Åke Lindahl, VP Environmental Affairs, Boliden Limited, November 1998

Mining companies have many priorities which compete for attention and resources. How important is APELL in the scheme of things for companies? Individual companies must make their own decision about that, taking into consideration that accidents do happen with potentially disastrous consequences; that a mine may be part of a community for a generation or more and is often a dominant presence in that community; and that the industry has a trust and a knowledge deficit. These considerations should make APELL an option worth pursuing.

Section 2

Relevance of APELL to the mining industry • Mining accidents and emergency preparedness • Existing policies, requirements and related activities by companies

Relevance of APELL to the mining industry

Like all industrial sectors, the mining industry has operational risks. Some are common with those in other industries that handle, transport and use chemical substances. Other risks are specific to the industry, particularly the storage of large quantities of tailings for extended periods. The physical risks are greatly exacerbated if tailings also contain high levels of bio-available metals or hazardous chemicals such as cyanide.

Mines and mineral processing facilities are often large, multifaceted operations, that have very specific interactions with the environment and with communities. They are found in remote locations or adjacent to towns and rural communities, in pristine or degraded environments, in areas of significant heritage or biodiversity, in all climatic zones, and in geomorphologically diverse areas. Accident risks and accident impacts vary greatly from operation to operation, probably more so than for industrial facilities which may be replicated at different sites in terms of process, risks and responses. Arguably this makes the whole issue of risk identification and emergency preparedness at mine sites more important than for some other industries. Each situation requires thoughtful, thorough assessment and planning. Mining operations and mining communities are also dynamic which means that plans need to be reviewed, tested and re-communicated.

As mine sites frequently occupy large land areas, there is often a considerable buffer between the mine and 'adjacent' communities. However, the apparent protection offered by remoteness needs to be tested by risk assessment. The large quantities of materials stored at sites or transported, the potential for spills to travel long distances downstream or to flow down slopes with dramatic speed, mean that there is no room for complacency even though an operation may not be in a highly populated area.

Many major mining accidents with off-site impacts have involved tailings management systems (dams, pipelines, etc.). There has been a smaller but significant number of chemical spills during transportation. Ground subsidence can also be a problem, particularly in historic mining areas. The industry has experienced other significant accidents such as explosives accidents although these are not known to have impacted adjacent communities. The industry can learn from experience in other sectors, and vision and vigilance must be watchwords. Learning from the experience of accidents, near misses and other industries is only common sense.

Table 2.1 describes the sort of accidents that can occur in the industry. Table 2.2 includes a number of significant accidents which have occurred.

Fuller descriptions of the risks present at mining operations are given in Sections 5 and 6.

Table 2.1: Potential accidents associated with mine sites and their effects

Type of Incident	Typical Causes	Potential Effects
Tailings dam failure	Poor water management, overtopping, foundation failure, drainage failure, piping, erosion, earthquake.	Loss of life, contamination of water supplies, destruction of aquatic habitat and loss of crops and contamination of farmland, threat to protected habitat and biodiversity and loss of livelihood.
Failure of waste rock dump	Instability often related to presence of water (springs, poor dump drainage).	Loss of life, injuries, destruction of property, damage to ecosystems and farmland.
Pipeline failure, e.g. tailings, leach solution	Inadequate maintenance, failure of equipment, physical damage to pipeline.	Contamination of soil, water, effects on water users. May not be detected for some time.
Transport of chemicals to/ from site	Inadequate transport procedures and equipment, unsafe packaging, high-risk transportation routes.	Contamination of soil, water, effects on water users, aquatic ecosystem damage, threat to human health.
Ground subsidence	Slope failure, breakthrough to surface.	Loss of life, damage to property.
Spills of chemicals at site, e.g. fuel tank rupture, reagent store damage	Poor maintenance, inadequate containment.	Contamination of soil and water. Air pollution could have health effects.
Fire	Poor design, unsafe practices in relation to flammable materials.	Effects of air pollution on health, property damage.
Atmospheric releases	Inadequate design, failure to follow procedures, inadequate maintenance.	Community concern, possible health effects.
Explosions (plant)	Inadequate design, failure to follow procedures, inadequate maintenance.	Community concern, loss of life, destruction of property.
Blasting and explosives accidents	Poor practice, unsafe storage and handling.	Property damage, risk to life.

Table 2.2: Examples of accidents that have occurred at mine sites

Accident	Impacts
Tailings Dam Failures	
Stava, Italy 1985	Tailings dam failure. 269 deaths in townships of Stava and Tesero, 7–8 km downstream.
Merrespruit, South Africa 1994	Tailings dam failure. 17 deaths in a community close to the dam.
Omai, Guyana 1995	Leakage from dam resulting in water containing cyanide entering the river system. Minor fish kills. Extensive discolouration of river for many kilometres.
Marcopper, Philippines 1996	Loss of tailings though old drainage tunnel. Evacuation of 1200 people, social dislocation of 700 families, damage to river systems and crops.
Aznalcóllar, Spain 1998	Dam failure resulting in loss of tailings and water containing heavy metals into river system. Farm land, crops and wells affected. Doñana National Park and World Heritage site threatened but not affected.
Baia Mare, Romania 2000	Spill of cyanide rich waters from tailings dam into river system. Extensive fish kills and economic impact on downstream communities. Contamination travelled through a number of countries, still detectable 2000 km downstream at mouth of the Danube.
Hau Xi Zinc Mine, China 2000	Failure of tailings impoundment. 15 deaths, 50 injured and over 100 missing as water and tailings engulfed houses and workers' dormitories.
Transport Accidents	
Kumtor, Kyrgyzstan 1998	Truck carrying cyanide overturned en-route to site. Perceived river pollution.
Tolukuma, Papua New Guinea 2000	Cyanide dropped from helicopter, near to a stream in remote area.
Yanacocha, Peru 2000	Truck carrying mercury lost part of load. 200–300 people suffered ill effects after collecting the mercury in the belief that it contained gold.
Waste Dumps	
Aberfan, South Wales 1966	Coal dump failed and engulfed part of the local town. 144 deaths.
Grasberg, Indonesia 2000	400 m high waste rock dump with its base in a lake failed. 4 contractors killed by water wave. Minor impact on downstream communities.
Subsidence	
Lassing, Austria 1999	Inrush of water and mud in underground mine trapped worker. 10 deaths during subsequent rescue effort. Subsidence resulted in crater at surface requiring relocation of families from threatened houses.

What are companies already doing which overlaps with APELL?

In many cases, companies will already be doing much of the work required to prepare for or to undertake a successful APELL process. Some governments already require such actions and more will do so in the future. Voluntary Codes are also likely to move in this direction. Companies are undertaking many related activities, such as risk management, community consultation and communication, social and environmental impact assessment and so on, which means that much of the information and many of the relationships required in the APELL process already exist. Existing on-site emergency preparedness planning can be extended to cover the off-site impacts of site emergencies. Planning for transport emergencies may require more attention than it has had in the past.

Existing requirements, policies and practices
Many government permitting procedures around the world require that a new facility should have an emergency response plan in place. Often it is part of the environmental impact assessment requirements. At present, few procedures specify involvement of the community in the plan formulation phase.

Legal requirements are moving towards more explicit consideration of community impacts and consultation on emergency plans. The European Commission's Seveso II Directive is an example.

(19) '....persons likely to be affected by a major accident should be given information sufficient to inform them of the correct action to be taken in that event'

(21) '..the public must be consulted on the external emergency plan'

Council Directive 96/82/EC on the control of major-accident hazards involving dangerous substances

The Australian Minerals Industry Code for Environmental Management is an example of the industry paying greater attention to contingency planning and to community consultation.

'Consulting with the community on the environmental consequences of our activities ...

Anticipating and responding to community concerns, aspirations and values regarding our activities ...

Applying risk management techniques on a site specific basis ...

Developing contingency plans to address any residual risk ...

Identifying interested parties and their information needs ...'

Australian Minerals Industry Code for Environmental Management 2000

Environmental Management and Auditing procedures such as ISO 14001 require that emergency preparedness and response measures are in place.

'4.4.7 The organization shall establish and maintain procedures to identify potential for and respond to accidents and emergency situations, and for preventing and mitigating the environmental impacts that may be associated with them.

The organization shall review and revise, where necessary, its emergency preparedness and response procedures, in particular, after the occurrence of accidents or emergency situations.

The organization shall also periodically test such procedures where practicable'

ISO 14001:1996 Environmental Management Systems—Specification with guidance for use

It is not clear to what extent mining companies are already carrying out emergency response planning in relation to off-site impacts and in consultation with communities. Many of the major mining companies (but by no means all) include emergency preparedness in their corporate policies. These high level policies do not focus on consulting with the community in the development of emergency response plans, nor on integrated planning with other agencies. More detailed company practice documents may do so. Boliden's Environmental Manual is an example.

As already mentioned, companies are unlikely to be embarking on a completely new process when they decide to implement an APELL programme. For many of them much of the information will be readily available and some of the procedures and community mechanisms will already be in place. However, some additional parties may need to be

Risk management and emergency preparedness

Why?

To minimize exposure to environmental risk the first step is the identification of risks. Once identified, risks can be evaluated and managed.

In the case of an emergency, it is of vital importance that the organization responds in a professional and efficient way. This will help to minimize negative environmental impacts, limit liabilities and help to protect the company credibility.

What?

Each operation shall:

- identify, evaluate and document on a periodic basis; potential environmental hazards and associated risks, and work to minimize these risks through hazard elimination, engineering controls, procedures, and education;
- prepare and maintain emergency preparedness plans which address potential emergencies on or around the site;
- identify and prepare for the management of reporting requirements in case of an emergency;
- encourage all employees to bring potential problems and risks to the attention of management.

Risk assessment shall be carried out in a formalized way. For some operations specific legal requirements apply and must be followed and documented.

Operations shall provide information to employees, other people on site (such as contractors, regular suppliers etc.) and interested people in the surrounding area and community about materials handled on site, processes and equipment related to known hazards and associated risks and the procedures for their control. Further, operations shall respond to community and public concerns.

Emergency preparedness plans should be developed in conjunction with emergency services, relevant authorities and the local communities.

A specified individual shall be assigned the responsibility for the development, implementation and periodic review of the emergency preparedness plan.

An organizational structure shall be established to quickly and efficiently direct and carry out emergency response activities.

Environmental Manual, Boliden, March 2000

contacted and specific tasks completed with regard to an integrated response plan. Relevant activities include the following:

- Emergency response plans and training to deal with on-site incidents such as spills, fires and accidents within the site boundary should be standard industry practice. All operations which subscribe to ISO 14001 will include emergency preparedness arrangements as part of this process. Auditing procedures typically require these response plans to be tested at regular intervals.

- For many new mines, community participation and consultation will have occurred as part of the Environmental Impact Assessment process, sometimes also with a separate Social Impact Assessment. Companies will know who their communities are, including their diversity and complexity.

- Some mining companies' social and environmental policies require on-going community consultation to take place at each site. Liaison committees comprising members of the local community and mine staff, which meet on a regular basis to discuss issues and concerns and operational matters, may be in place. Emergency preparedness could be added to the agenda of such a group or delegated to another group.

- Mining companies will have extensive contacts with government agencies, local and national, with EPAs, mines inspectorates, etc. which may be part of the emergency response process. They should also have established links to emergency response agencies as a result of planning for on-site accidents.

- The Closure Planning process necessarily involves close consultation with local communities. Closure planning is an ongoing and dynamic activity, reviewed at different stages during the mine life and finalized towards actual closure in the light of changing needs and aspirations of the local community. Long-term environmental stability and safety of the site is a central issue in closure planning.

- Various risks at mines may be reviewed regularly for insurance purposes. Underwriters will be looking for risks such as loss of production, worker claims, etc. but also off-site third party claims. The extent of the area that may be vulnerable in the event of an accident and the number of people that may be affected may already have been assessed for these purposes, and this information could also be used in the emergency preparedness planning process.

- Occupational health and safety audits, and engineering safety checks may also include procedures that can be used as a basis for collecting information on potential hazards for the emergency preparedness planning process. Information on hazardous chemicals, such as that contained in Material Safety Data Sheets (MSDS), will already be required by health and safety procedures.

Section 3

Preliminary actions • Raising awareness and gaining commitment • The Co-ordinating Group and its composition • Defining the local community • Case study: East Malartic Mill and the town of Malartic

Any stakeholder group can decide that the present state of emergency preparedness at the local level is inadequate and that there is a *prima facie* case for reviewing the state of planning to make improvements where necessary. Companies are likely to be the initiators in most cases.

Preliminary actions

Before starting on the steps of APELL, some preliminary actions can help the process to get underway smoothly. There must be sufficient communication about emergency response issues and about the APELL process in advance of getting started to build a degree of support and enthusiasm so that people are prepared to commit their time and effort, and that of their organizations, to the work which will be required.

Suggested actions to help in getting started are:

- Undertake an initial stock-take of the obvious hazardous aspects and risks of the operation and of emergency procedures in place.
- Identify, in a preliminary way, vulnerable communities and key individuals/agencies/organizations which should be involved.
- Develop familiarity with the APELL process.
- Raise awareness and gain commitment through internal company seminars and community workshops.
- Establish an informal Co-ordinating Group to get planning and communication underway.

Raising awareness and gaining commitment

Further thoughts on raising awareness and gaining commitment are given below.

Running an internal company seminar

An internal company seminar may be the best way to ensure there is adequate management understanding of the need, benefits and risks of launching an APELL process and to ensure that resources are available to do it properly.

The seminar would cover, in a preliminary way, such things as:

- the operation's hazards and potential risks
- some accident scenarios and their potential consequences
- regulatory or Code requirements
- vulnerable communities
- adequacy of current plans and
- the APELL process

It could usefully review relevant information and input which the operation already has which will help with the process, identify gaps and assign accountability for the work.

Raising awareness and gaining commitment in the community

The amount of preparatory awareness-raising required will differ in each case. In many places it appears that public knowledge of mining and its associated risks for the community is low. Indeed, after accidents have occurred, it has frequently emerged that the community did not understand either the operation or its risks. The community awareness pre-step is therefore important, not only as a precursor to getting commitment to the APELL process, but in its own right. In the case of mines with local indigenous populations, the challenge of communicating effectively can be compounded, particularly in the development stages of a mine, when traditional communities may have no real appreciation of what is involved.

APELL Workshop

Experience shows that running an APELL Workshop involving managers, community and government representatives, and emergency response providers can be an effective and worthwhile early activity. It helps to promote enthusiasm for the project, can be used to identify additional members for the Co-ordinating Group, present a broad view of the current status of emergency planning, generate some useful data for the planning process and initiate contacts with the media.

Two days may be required for such workshops, which in fact go beyond awareness raising and gathering of ideas, and start on the process itself. The APELL process should be explained. It is useful to have presentations on accident related topics including experiences, and lessons learned. Each representative participant should present themselves and their organization and their role in a potential emergency. The initial strengths and weaknesses of the preparedness of the community may be assessed and suggestions for improvements discussed. Ideas for simple risk reduction may be generated. A start can be made on developing a project plan with proposals for next steps.

Once the Workshop is over it is important to maintain the momentum. Preliminary proposals and ideas from the Workshop would form the basis for discussion by the Co-ordinating Group.

The Co-ordinating Group should be finalized with individuals identified at the Workshop added or its composition confirmed. The remainder of the APELL process can begin, although a good deal of the data and much of the thinking required will already have been generated in these early stages and the work will be off to a good start.

'In our experience, formation of the APELL Co-ordinating Group, after some initial discussions, is the first important step which should precede all other APELL activities. The Group is first set up informally and then either formalized or expanded by organizing the APELL Workshop. The local authorities and the managements of industries (in this case mining) are well aware of the names of reputed community representatives who can be invited to be members on APELL. Building community awareness is an on-going and multi-activity sub-programme, which can only be undertaken if the APELL Co-ordinating Group is in place and functioning well.'

K C Gupta, National Safety Council, India

Awareness raising by governments or industry associations

Over 30 countries have used the APELL process to catalyze action in their countries, or in particular regions. The APELL Workshop is typically organized by governments to alert industry, communities and emergency providers alike to the benefits of better planning and communication and to provide an introduction to the APELL tool to assist in specific planning.

Industry associations should be proactive in tackling this issue with their members. They can run seminars to promote the importance of emergency preparedness and to inform their members about the APELL concepts and process. They can play a constructive supportive role in the broader, community-based APELL Workshop. Industry associations are also increasingly expected to provide information, technical background and comment in the event of a major accident. They need to be proactive in making their own preparations.

Checklist: inputs to awareness raising

- define the local community(ies) concerned;
- list existing local community contacts;
- identify other mines or industrial facilities to be involved;
- gather information on existing emergency services and community response plans;
- prepare presentation materials on the mining operation, its hazards, and existing emergency response plans;
- select appropriate methods of communication;
- develop an introductory presentation on APELL, its benefits and requirements;
- form an informal Co-ordinating Group to plan the initial consultation processes, including possibly a town meeting, a seminar, an APELL Workshop.

The Co-ordinating Group

The Co-ordinating Group has already been introduced as one of the integral elements of APELL, along with its Ten Steps. This Section goes into more detail about the roles and composition of the Group, and matters to be considered in order to maximize its effectiveness.

The Co-ordinating Group provides oversight and energy, and brings together the views of respective players to drive the APELL process and to ensure it is inclusive and effective. The Co-ordinating Group does not have a direct operational role during an emergency. It has a key role in building and maintaining motivation, communication, commitment, cooperation and momentum during the project. More specifically, the Group's roles include:

- ensuring open lines of communication between all parties;
- identifying key people and organizations to involve;
- setting objectives and a timeline for the process;
- overseeing development of the coordinated emergency action plan (10 Steps);
- identifying available expertise;
- establishing working groups for specific tasks;
- ensuring clear risk communication occurs to vulnerable communities;
- preparing the various parties involved to know their tasks should an accident occur; and
- remaining as a central forum for dialogue and review after the planning process is complete.

Composition of the Co-ordinating Group

The Co-ordinating Group should include representatives of those parties who are responsible for minimizing and for responding to emergencies, or who have a legitimate interest in the choices among planning alternatives.

Mine managers, emergency response providers, environmental agency representatives, and community leaders would form the core. Identify people with a wide range of relevant expertise and local knowledge. Imagining a worst case scenario can identify who would be affected by an accident. The Co-ordinating Group and the emergency response participants who are identified in Step 1 of the APELL process are not necessarily the same. There will be some overlap, but the Co-ordinating Group will include other interested participants who have no particular response role in an emergency.

The Group cannot include everybody and would quickly become unwieldy if allowed to expand too much. An effective working size should be maintained. If there is widespread interest and demand, a broader consultative conference could be established.

Selecting effective people is, of course, more important than having the right organization represented by the wrong person. Individuals need the right personal attributes and must have the respect of their constituencies as well as commitment to the process. They must be able to cooperate with one another during development of the plan and after it has been developed to ensure that there is no loss of preparedness when changes occur in the local area (e.g. new industrial facilities, new housing developments, etc.).

Additional considerations relevant to the composition of the Co-ordinating Group are:

- In the case of remote mine sites, very few people may be affected, or the authorities responsible for nature conservation, catchment management, etc. may be based some distance away from the mine site, making it impractical to include a representative in the Co-ordinating Group. Communication with these agencies would nevertheless be required at different stages in the process, as well as about the Emergency Response Plan itself.

- Cultural and political diversity should be considered when setting up the APELL Group, e.g. the composition of the local population may differ from formal governance structures.

- There may be a large local population which has very little in the way of organizations, but where village elders and headmen would be important members of the Group.

- Officials from some government agencies who are non-resident may be appropriate members of the Group, but the frequency with which they visit the area would be a factor.

- Consider including a member of the local media in the Co-ordinating Group, whether or not they are likely to have a direct emergency response role.

A check-list of people or organizations to be considered for participation is set out below.

Individuals for the Co-ordinating Group could come from a wide range of organizations

- mine managers, HSE, external affairs, communications staff
- mines inspectorate representatives
- chemical suppliers
- transport operators—road (public and private), rail, air, water
- members of environmental or other non-government organizations
- representatives of Natural Resource agencies
- member of the Local Planning Authority
- representatives of agencies with responsibility for fire, health, water quality, air quality, emergency response planning
- representatives of other major facilities in the area
- local hospital/medical representatives
- teachers and community education representatives
- representatives of private sector, such as trade organizations, industry officials
- representatives of labour organizations
- members of Community Councils
- members of residents associations
- representatives of heritage organizations
- civil defence chief
- National Parks officers
- local religious leaders
- translators, if more than one language is involved

Leading the Co-ordinating Group

Effective leadership of the Co-ordinating Group can be a critical factor in ensuring a smooth, co-operative and productive project.

The Leader of the Co-ordinating Group should be able to motivate and ensure cooperation of all of those involved regardless of cultural, educational, or institutional status and other differences. The degree of respect held for the person by the other members of the group will be important, but so too will the very practical considerations of availability of the person's time and resources, their experience in managing group relationships, and their skills in cross-cultural communication.

Although the company may have initiated the process and be vitally interested in doing it well and efficiently, it does not necessarily follow that the mine manager will be the best person to take the role of Group Leader. It may be better for a member of the community to take the role. If the mine manager is an expatriate, cultural differences may be an impediment. Also, continuity with a resident population should be a leadership consideration.

In some cases it may be appropriate to share the leadership of the group between an industry representative and a local authority member, taking into account both personal and institutional considerations.

As well as ensuring that the Co-ordinating Group fulfils its roles as described above—particularly establishing clear objectives, timing and resources for various phases of the process—the leader should ensure that the process remains a cooperative exercise rather than a place for negotiation and that the APELL agenda does not become embroiled in other unrelated issues. The community and its preparedness in case of accident remains paramount.

Defining the local community

In the APELL process, thought needs to be given to which communities may be affected by an accident. The mine may already have extensive relationships with local community groups but they may not represent all who could be affected during an accident (e.g. a transport accident), nor all those likely to have a legitimate interest in accident risks and responses.

In defining the community, there are some obvious practical starting points related to specific accidents which may occur. The affected community in the case of an accident may not be the same as for day-to-day mining issues. The vulnerable community will depend on the type of hazard and an estimation of the worst credible accident.

What areas downstream of the mine site and its facilities such as tailings dams may be affected by a catastrophic spill? More than one catchment may be affected and people living many miles from the site could potentially be affected. What is the prevailing wind direction and what communities lie down-wind? What are the chances of less frequent wind directions affecting other populations who should also be included? What about deliveries of hazardous materials to the site? Are they considered from the time they leave the manufacturer to the time they arrive at the plant

The following may be considered in defining the community:

- geographic or administrative boundaries
- catchment boundaries (airshed and watershed)
- governing bodies affecting the operations
- traditional landowners
- influential organisations such as civic, religious, educational, etc.
- major media
- concerns of local residents

site, or is there some other convenient cut-off point? Are the hauliers of the hazardous materials in contact with communities along the route?

When mines are developed from in-house exploration, contacts, relationships and knowledge of communities, individuals and attitudes may have been built up over many years. It may prove difficult for companies to maintain that knowledge and to sustain personal relationships in the transition years through construction and operation. As noted earlier, new mines may also have in-depth baseline social impact studies to draw on, and may have conducted community consultations as part of their planning and approval processes. Even so, there may be difficulties in identifying all of the right people with whom to consult, and in sustaining the process over many years to keep plans and knowledge current.

Other considerations

Migratory populations and Native Title Holders may be affected

In several parts of the world, migratory populations such as the Bedouin in the Middle East, First Nation populations in North America, Asian herdsmen and other indigenous peoples may form part of the community influenced by the mine. Their settlements, migration routes, grazing pastures, hunting grounds or sacred sites may be affected as a result of a mine accident, but they may only be present in the vicinity of the site for a few weeks a year. While the direct involvement of these groups in the APELL process could be impractical, their interests should be considered and there may be representatives who can be included. Native Title Holders should also be included. They may live in the vicinity of the mine—permanently or periodically—or may visit their land regularly.

Induced development

In many parts of the developing world, induced development or the 'honey pot' effect of mines is a feature. When a development that provides jobs occurs, the local population may increase greatly as people move to the area to find employment. This is often the situation with mines in remote locations and can dramatically alter the risk scenarios in certain locations as well as the risk-weighted consequences of an accident. The composition of the Co-ordinating Group as well as the communication strategies may need to be reviewed in a long-running APELL process if the demographics of the local community change significantly.

How the getting started steps fit together

These pre-APELL steps can be seen as interrelated and mutually supportive. For example, the early formation of an informal Co-ordinating Group would enable more effective planning for the awareness-raising steps as well as proper planning for the APELL Workshop. Similarly, the Workshop will play a large role in the awareness-raising programme and will help to crystallize the composition of the Co-ordinating Group.

None of these steps should be made unduly onerous. They need to be kept in perspective as sound but proportionate preparation for the processes to follow.

Getting Started—Case Study of Sulphur Dioxide Safety at the Barrick Gold Corporation, Bousquet Complex, East Malartic Mill and the Town of Malartic.

The process

This case study describes the process initiated by the management of the Barrick Gold Corporation, Bousquet Complex, East Malartic Mill to prepare and protect its employees in the unlikely event of an accidental airborne release of sulphur dioxide at the mill and to assist the Town of Malartic to do the same for its citizens. The process included the definition and minimization of risk, and updating of emergency preparedness and response plans and capabilities for both the East Malartic Mill and the Town of Malartic. The ultimate success of the process was predicated on a high level of involvement by all parties concerned, not the least of which are the citizens of Malartic.

The context

The East Malartic Mill is located within Malartic, a town of approximately 4000 inhabitants, in Quebec, Canada. Within the mill, cyanide is used to recover gold from ore that is trucked from the Bousquet Mine approximately 35 km to the west. Before being pumped from the mill, the slurry containing the spent ore is treated to destroy cyanide in a process that utilizes sulphur dioxide.

Risk definition

Sulphur dioxide is toxic, corrosive and a strong irritant. At the East Malartic Mill, it is received and stored under pressure in liquid form. In the event of an accidental release, it would volatilize and could create a ground-hugging cloud that could form sulphuric acid upon contact with moist skin and mucous membranes of the eyes, nose, throat and lungs.

A detailed risk assessment was undertaken by a specialist consulting firm to better define the risks related to the use of sulphur dioxide at the mill. This assessment yielded two principal conclusions. The first was that, even though the liquid sulphur dioxide storage and handling facilities were judged to be very good and the probability of an accidental release was judged to be low, improvements could be made to further reduce and manage the risk. The second was that, in the unlikely event of an accidental release and unfavourable atmospheric conditions, the citizens of Malartic as well as personnel working at the mill could be affected.

Emergency preparedness and response—East Malartic Mill

In the light of the risk assessment, steps were taken at the mill to reduce risk, and improve emergency preparedness and response. These included: improving protection of the facility against impacts by vehicles; increasing the frequency of inspections and preventive maintenance; installation of pressure indicators and leak detectors to provide early warning of an accidental release; installation of automatic, remotely activated valves to prevent accidental releases; installation of an alarm, with a sound distinctly different from that of the fire alarm, to warn employees in the event of an accidental release; placement throughout the mill site of personal protective equipment that would facilitate employee evacuation; retraining of all employees regarding emergency evacuation procedures; updating of emergency response procedures and equipment; and retraining of emergency responders.

continued ...

Emergency preparedness and response—Town of Malartic

After discussions with other companies that also store and handle liquid sulphur dioxide and concurrent with efforts to improve emergency preparedness and response at the mill, contact was made with representatives of the Quebec Ministry of Health and Social Services. This proved to be beneficial, in that the Ministry reviewed the risk assessment that had been completed and provided feedback. In addition, the Ministry also advised that its representatives, together with representatives of the Quebec Ministry of Public Safety, had met previously with representatives of the Town of Malartic to discuss the risks related to the transportation and storage of hazardous materials, including sulphur dioxide, within the town limits. The principal conclusions of this meeting were that the Town of Malartic was responsible for the safety of its citizens, and that the town should modify its emergency preparedness and response plan to address the possibility of an accidental release of sulphur dioxide at the mill. It was also concluded that, in order to revise its emergency preparedness and response plan, the town would need to collaborate closely with the management of the mill.

After having contacted Ministry of Health and Social Services representatives, the management of the mill took the initiative to organise a meeting with the Health and Social Services, and Public Safety Ministries and Town of Malartic representatives. At that meeting, mill management provided an overview of mill operation, and presented details of the risk assessment and the risk reduction measures that had been implemented. The overview of the operation included a mill tour as well as descriptions of the cyanide destruction process and sulphur dioxide handling, storage and use. Ministry of Health and Social Services representatives provided details of the characteristics of sulphur dioxide and its potential health impacts. Ministry of Public Safety representatives emphasized the need for town and mill representatives to jointly develop communication, and emergency preparedness and response plans. As a result of the meeting, town management and mill pledged to work together to develop the necessary plans.

In the near future, the town intends to form committees to finalize its communication, and emergency preparedness and response plans. These committees will include representatives of the mill, the town, public service organizations (e.g. police, fire, ambulance, hospital, public works, schools, etc.) and the general public. The purpose of these committees will be to develop inventories of the available and required response resources (personnel, equipment and systems), and to define the response actions to be taken by each entity (East Malartic Mill, Town of Malartic, public service organizations and general public). In addition, the committees will decide the method to be used to inform citizens of an accidental release; the emergency action to be taken by citizens; and the criteria for establishing that the danger has passed, as well as the method of advising the citizens. The committees will also plan and conduct the initial sessions at which citizens would be informed of the potential risks, of how they would be notified in the event of an accidental release and of the actions that they would need to take to avoid being affected. Updates of this information would be provided to the citizens on a periodic basis via mailings, newspaper articles and television/radio messages. Tests of the emergency notification system(s) would also be undertaken at the time of the initial information sessions and on a periodic basis thereafter to ensure a proper level of preparedness and to maintain public awareness.

The activity described above, which is scheduled to be completed in 2001, will conform fully to recently proposed Quebec legislation that will require those responsible to declare the risks that are inherent to their activities to the municipalities in which they operate and to implement, together with civil security authorities, appropriate monitoring and alert procedures. In addition, the legislation will establish criteria for municipalities for informing their citizens and for the development of emergency preparedness and response plans. Other sections of the legislation will address the declaration of a state of emergency, the need for mutual aid agreements and the responsibilities of risk generators, individuals, municipalities and government.

Conclusion

The objective of the process initiated by the management of the East Malartic Mill was to better prepare and protect its employees in the unlikely event of an accidental sulphur dioxide release and to assist the Town of Malartic to do the same for its citizens. Achieving this objective involved providing all parties with a good understanding of the risks related to an accidental release of sulphur dioxide from the mill and ensuring that, in the unlikely event of an accidental release, plans are in place to ensure rapid and effective response by all parties, including the citizens.

Authors

Christian Pichette, Manager, Bousquet Complex

Dominique Beaudry, Environmental Co-ordinator, Bousquet Complex

Pierre Pelletier, Superintendent, East Malartic Mill

Section 4

The Ten Steps of the APELL process
* ***Communications in an emergency***

The Ten Steps of the APELL process

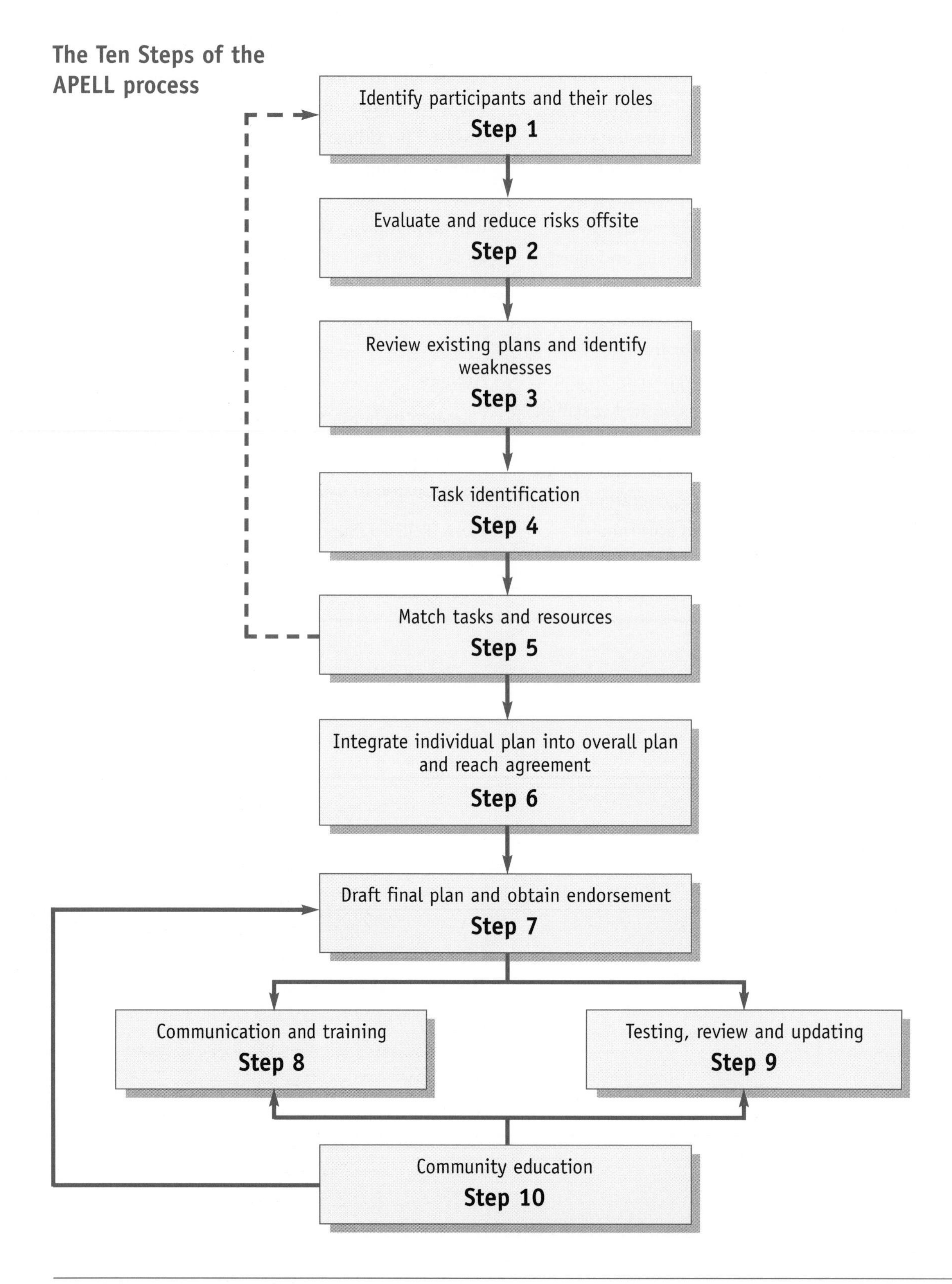

Step 1—Identify emergency response participants and establish their roles, resources and concerns

Members of the Co-ordinating Group will be well placed to understand the emergency response agencies and the resources available in the local area or will know where to get the information.

The following tasks are involved in Step 1:

- Compile a list of potential emergency response participants. See box below for list of possible providers. In addition, Co-ordinating Group members may be aware of specialist groups which could be called on in specific emergency situations.
- Obtain copies of existing emergency plans and review these to identify any further emergency response agencies and participants.
- Establish concerns, e.g. about deficiencies in resources or weaknesses in response capabilities.
- Prepare a brief description (perhaps a spreadsheet) of all emergency participants, their roles and resources, e.g. personnel, equipment, special knowledge, facilities, etc. Pay particular attention to understanding and documenting the boundaries between the different providers, gaps, overlaps and any unclear roles and responsibilities.

Checklist of emergency response participants:

For mines operating in remote areas, many of these agencies will not be present and company personnel will fulfil many of these functions.

- fire department
- police
- emergency health services such as ambulances, paramedic teams, poisons centres
- hospitals, both local and for evacuation for specialist care
- public health authorities
- environmental agencies, especially those responsible for air, water and waste issues
- other industrial facilities in the locality with emergency response facilities
- civil defence teams
- welfare services
- Red Cross/Crescent
- public works and highways departments, port and airport authorities
- public information authorities and media organizations

Mining-specific issues to consider

Many mines operate in remote areas where government agencies and infrastructure may be extremely limited or, in places where they exist, they may be severely under-resourced. In such cases the mine will provide virtually all of the resources necessary to deal with emergencies. It may already have provided equipment e.g. community ambulances, or training to local groups such as volunteer fire-fighters.

In some cases, towns or settlements adjacent to mines may have grown substantially as a result of the mine's presence, and outgrown the capabilities and resources of its emergency response agencies. Volunteer organizations may exist that are capable of fulfilling a role, e.g. organizing an evacuation.

Some mines are fly-in fly-out operations remote from towns and from emergency response agencies. Some are in pristine areas with no local community, where the emergency response would be aimed at preventing damage to sensitive ecosystems. Again, the mine may have to provide most of the equipment and facilities to be able to react to an incident. However, there may be agencies responsible for or NGO's concerned with the protected areas that may also be able to mobilize staff and equipment in the case of an accident.

In other cases, mines are located in highly developed areas with efficient emergency services and environmental agencies present. There is, therefore, a wide variety of situations and the inventory of potential emergency response providers and available resources will be different in each case.

Step 2—Evaluate the risks and hazards that may result in emergency situations in the community and define options for risk reduction

Possible accidents should be identified, along with the probability of their occurrence and possible consequences. This enables scenarios to be constructed and priorities to be set for planning purposes. Simultaneously, apparent risk reduction options should be defined and pursued.

- The Co-ordinating Group should oversee the compilation of a list of hazards and potential risks. Work must be done to explore and comprehend the range of hazards which exist, in addition to focussing on the obvious. To assist, consider:
  - reviewing mining accidents which have occurred, including near misses, or incidents that similar facilities have experienced. (Refer to Tables 2.1 and 2.2 for types of accident. Also refer to descriptions of Mining Hazards in Section 5 and Accident Case Studies in Section 6.);
  - the experience of chemical or transport accidents in other industries since mining operations handle hazardous materials of a general nature;
  - natural disasters such as earthquake and floods and forest fires, which may cause or compound an emergency at an operation;
  - seasonal hazards—freezing may contribute to the occurrence of some accidents, the spring thaw will contribute to others. Some accidents may be more prone to occur in the dry or wet season in parts of the world;
  - also consider the community's perception of risks and its willingness to accept certain risks but not others. This dimension is important and risk assessment can usefully be approached as more than an engineering, technical exercise.

- Assess the potential severity of the impact, for each possible accident, e.g.:
  - the size and nature of potential area affected
  - number of people at risk
  - type of risk (physical harm, toxic, chronic)
  - long term effects
  - impacts on environmentally sensitive areas
  - consequential secondary risks and impacts

- The probability of occurrence should be assessed, either qualitatively or using a quantitative assessment. Points to consider include:
  - probability of individual events
  - probability of simultaneous events (e.g. earthquake resulting in rupture of a pipeline)
  - complications from unique environmental considerations, e.g. severe terrain, location on a major river, frozen conditions, etc.

- The Co-ordinating Group should agree on key scenarios that could reasonably be expected to occur or that the community is most concerned about and use these in the planning process.

- As the hazards are identified and their probability and consequences are examined, some areas of risk may be identified that can be readily eliminated or cost-effectively pursued. Appropriate action should be taken to reduce or manage those risks through changing operating practices, upgrading equipment, training, changing chemicals used, etc. The emergency planning process complements but does not substitute for risk management and risk reduction—action must also be taken.

A specialist team or other group may be required to recommend risk reduction options rather than the Co-ordinating Group, but results, plans and progress should be reported back to the Group. It may be possible to completely eliminate some risks. If this is the case, this can be documented and the remainder of the APELL process can concentrate on the remaining risks.

Risk assessment and risk management—defining the terms

'To assess risks, the hazards must first be identified. A Hazard is a property or situation that in particular circumstances could lead to harm. Consequences are the adverse effects or harm as a result of realizing a hazard, which cause the quality of human health or the environment to be impaired in the short or longer term. Risk is a combination of the frequency of occurrence of a defined hazard and the magnitude of the consequences of the occurrence.'

Royal Society 1992

'Risk management involves using the information from risk assessment to make and implement decisions about risk based on the balance between costs and benefits, for a range of options that deliver the intended course of action. Communication of the scale and elements of risk to those involved is a key part of a risk management strategy.'

HMSO 1995

'Risk management techniques attempt to guide decisions by a logical and systematic consideration of possible future outcomes rather than historical precedent. They encourage the consideration of both risks to the stakeholder and of the costs and benefits of activities associated with risk. In theory this should decrease the tendency to consider only short term and financial outcomes of decisions.

In practice, risk management methodologies have not yet fully come to grips with how to identify possible future scenarios. Here we still tend to fall back on using experience to predict what might happen in the future. The major challenge for risk management is to find better ways of considering possible future outcomes so that decisions can take into account a longer term view.

Risk management involves communication and consultation both internally and externally. A crisis situation is, to a large extent, an information and communication crisis where a key variable in determining vulnerability to a crisis is communication.

There are usually many factors other than the level of risk to take into account. Not the least of these is the opinions of stakeholders. Often risks are carried by people other than those who will directly reap the benefits. A level of risk imposed on others is usually expected to be lower than the level of risk accepted by people who experience the benefits. Misperceptions of risk are also important, and very difficult to deal with ... Communication and consultation is again critical.'

Risk Management and the Future,
Australian Minerals and Energy Environment Foundation 2000

Risk assessment in mining

The use of risk assessment has become commonplace in the mining industry. For example, risk assessment is routinely used in the design of engineered structures in the mining industry, including tailings dam walls. Computers analyse the response of a design to static (e.g. accumulation of tailings and raising the dam wall) and dynamic (earthquake or other shock) loadings, and a range of options for construction are assessed. Risks can be reduced by mitigation methods such as improving foundation strength and construction material specification. Abnormal conditions and contingencies should be identified, and engineering designs reviewed to ensure that the design allows for these situations.

For complex installations, a Process Hazard Analysis may be routinely used during design. This looks at the risks arising from the management of hazardous materials and processes in an industrial plant. Risks can be reduced by changes in plant layout or by improvements in the storage and handling of hazardous materials, and incorporated in a revised design.

Operating procedures need to cover situations such as start up of plant and equipment, which may involve different conditions with associated environmental risks.

Step 3—Have participants review their own emergency plan, including communications, for adequacy relative to a coordinated response

Emergency plans may exist in various forms for many areas, for example, regional and local plans, police and fire plans, hospital plans and mine site plans. National Disaster agencies or coordinators are one source of information on existing plans. A list of these is available from UNEP and the Office for the Coordination of Humanitarian Affairs (OCHA) (www.reliefweb.int/ocha). In some sparsely populated and remote areas, where mines operate, emergency plans may be completely absent. In others there may be unwritten responses which need to be understood in dealing with emergencies such as fire at the local level. The objective of this step is to review plans for adequacy in the context of their contribution to an effective overall response to the emergency scenarios agreed by the Co-ordinating Group.

- Contact the participants identified in Step 1, outline the priority emergency scenario(s) and ask them to evaluate their plans against these scenarios. A checklist of plan elements, response tasks and equipment can be developed to assist in this evaluation. Some key elements where deficiencies can arise are listed below. This checklist can be further developed by reference to the 'Components of an Emergency Response Plan' listed in Appendix 1.

- The Co-ordinating Group should review the results of the separate evaluations to determine the overall strengths and weaknesses of the current status of a coordinated emergency response. The checklist can be developed into a table to compile information on the various response plans prepared by different agencies. This table will help to highlight gaps in relation to an integrated emergency response plan.

Key elements of an emergency response plan

- communications equipment that can reach all participants, such as mobile phones, pagers, short wave radios depending on location
- media contacts and media relations strategy developed
- specialised hazard monitoring and training such as dealing with chemical fumes or water pollution
- adequate emergency equipment for spill containment or collection such as additional supplies of booms and absorbent materials
- clear reporting procedures
- clear procedures and alarm signals
- alerting the public and coordinating evacuation using sirens or other warnings; well rehearsed warnings, evacuation procedures and easily reached shelters
- role of the participants in different areas of response, such as fire-fighting, community protection
- alternative drinking water supplies in case usual supplies are contaminated
- rapid test kits for chemical spills, e.g. cyanide in the case of gold mines
- readily available access to information on dealing with chemical hazards
- options for clean-up following the accident scenarios examined. These could be both the immediate actions to be taken, and the approach which would be taken to a longer clean-up programme

Step 4—Identify the required response tasks not covered by existing plans

From the reviews carried out in Steps 2 and 3 it can be determined whether existing emergency plans adequately address the identified risks and emergency scenarios. Additional tasks that need to be undertaken to complete or improve the plan can be identified. This step requires a thorough definition of what more must be done, with input from emergency response participants and Co-ordinating Group members.

- Identify missing or weak elements or tasks not being covered by any group, in the context of an integrated response.
- Determine the importance of these elements to the function of the participant (e.g. the fire service may not have the proper equipment to fight some chemical fires; correct antidotes may not be available at nearby hospital).
- Inter-relationships, responsibilities and communication plans are key items for the Co-ordinating Group to discuss. For an effective integrated response, the importance of establishing a clear command structure cannot be overstated.

Step 5—Match tasks to resources available from the identified participants

Each task defined in Step 4 must be assigned by the Co-ordinating Group to the participant who can best address that aspect. Assigning the tasks should take into account authority, jurisdiction, expertise or resources.

- Evaluate each of the required extra tasks separately and, using the list of participants from Step 1, determine who is most likely to be able to complete the task. Assess benefits or problems associated with a particular participant completing a particular task.

The importance of overall command

A tanker containing fuel oil turns over on a busy road and the driver is trapped inside. The emergency services are called. The fire crew will take the initiative to remove the injured driver, and in reducing the risk of fire and explosion from the leaking tanker. The driver will be taken to hospital by ambulance. Meanwhile, fuel oil is spilling into a major river, upstream of an intake that supplies the nearby city with drinking water. The road is blocked, holding up traffic, so the police are also on the scene. Who is in overall charge? The police, ambulance, fire crew, and environmental response team who may be able to contain the spill that threatens the city are all present and must work together to minimize the damage that could result from the accident. APELL would have identified who would take command prior to such an event, to avoid confusion during an emergency. (Example adapted from TransAPELL)

Some areas have established multi-agency 'command posts' to resolve the issue of who is in charge.

Within the mining company there must also be a clear prior understanding of command responsibilities. Local management will obviously be intimately involved, but head offices and corporate staff would be expected to have a role in certain critical decisions. Valuable time can be lost during an emergency if consultation is required as to who is in overall charge for the company or as to the sorts of emergencies or decisions that are required to be taken at Head Office. It is important to distinguish in advance between an emergency that can and should be handled by local management in accordance with the Emergency Response Plan and one which requires corporate senior management involvement and action. Factors which may determine corporate involvement include:

- significant threat posed to the public
- significant government or media scrutiny (international or national)
- likelihood of escalation with no immediate resolution in sight
- significant threat to corporate reputation and loss of shareholder value

- Discuss the tasks with the participant to determine willingness to undertake it and their resources and experience that will ensure the task is completed, or identify problems which may make it inappropriate or difficult for them to do so.

- Determine if any new tasks, problems or constraints will arise as a consequence of completing those already identified.

- Monitor the successful completion of each task.

Resourcefulness and initiative may be required here. For example, in a sparsely populated area where police are scarce, volunteer fire fighters could be used for temporary traffic and access control.

Step 6—Make changes necessary to improve existing emergency plans, integrate them into an overall community plan and gain agreement

By completing Steps 4 and 5, all resource-related problems should be identified and resolved. Integrating the plans will reveal overlapping responsibilities and complex interfaces between agencies. Tasks at this Step are to:

- Prepare a draft integrated plan.

- Ensure that the newly developed plan is consistent with any regional disaster plans; also ensure its consistency with legislation and any codes which are relevant to emergency planning and community engagement.

- Check that the plan is robust in relation to all previously identified risks and emergency scenarios and in relation to response tasks, resources, roles and accountabilities, etc., to ensure there are no weak components.

- Conduct a role-playing exercise to test the plan, with key participants describing how they would respond to a variety of different emergency scenarios.

- Identify any weaknesses in the plan and, if necessary, repeat the two previous steps to resolve these problems.

- Revise the plan as often as necessary until all deficiencies are eliminated and the members of the Co-ordinating Group agree it is appropriate and workable.

- Ensure that any individual plans which the various providers and organizations may retain to focus their own particular responses are retrofitted to the integrated plan and that inconsistencies are not allowed to creep in.

Aiming for clear usable plans
Successful plans are often simple, supplemented with appendices of detailed information where necessary.

Many plans include a telephone and contact roster, an action guide/checklist, list of resources/capabilities that can be shared, and an action checklist for field use. Plans that fill thick files are unwieldy and more likely to be ignored or bypassed. Simple, uncluttered flow charts are easy to use and more people can be expected to use them in an emergency with no special training.

Some or all of this information can be made readily accessible on company websites, however the plan needs to be available in hard copy as well since computer systems may be disrupted by the accident itself or by other failures.

Post-emergency clean-up as part of the plan
As mentioned, the issue of clean-up after an emergency should be considered in the planning process, otherwise problems may be encountered later on. Collecting base-line data relevant to the risk scenarios is one important element. Another is to have considered in general terms the logistics, benefits and downsides of alternative clean-up and remediation strategies so that immediate action taken in the course of an emergency does not complicate the longer-term approaches to effective remediation. Clean-up operations can themselves be dangerous and risk assessments are advisable when considering the options.

Detailed clean-up and remediation plans can of course only be prepared after an accident has occurred. Apart from minimizing the environmental and social impacts of the accident itself, the objective would be to enable the mine to return rapidly to safe production with clean-up to standards acceptable to regulatory authorities, to the community and to the company itself, consistent with good corporate citizenship.

'With respect to the Aznalcóllar clean-up operation, one problem that we had to face was the fact that we had to start the clean-up without having any official clean-up criteria established. They were established after the clean-up was more or less finished. We dealt with this by establishing our own criteria based on internationally accepted levels combined with a site specific risk assessment looking at residual levels, exposure and potential effects'.

Lars-Åke Lindahl, VP Environmental Affairs, Boliden Limited

Step 7—Commit the integrated community plan to writing and obtain endorsement for it and relevant approvals

The integrated plan, as agreed by the Co-ordinating Group needs to be documented in final form and endorsed by the community, local government or other appropriate agencies.

- Use a small group to write the plan in its final format.
- Prepare a standard presentation to be given to the community, government officials or others who may have a role in its approval or implementation.
- Prepare notices, instructions, posters, etc. for use at the site and by other organizations and individuals.
- Make presentations, hold meetings and review sessions and obtain endorsement of community leaders and relevant officials.
- Make arrangements for any written agreements that may be necessary between participants of the APELL process, such as mutual aid, notification formats, use of the media, specialized response personnel and equipment. Agreements are also needed when private companies are to provide particular emergency assistance such as technical expertise or specialized equipment.

The objective of this step is to take the plan from the development stage, during which the APELL Co-ordinating Group has been the 'owner' of the emerging plan, and to transfer ownership and endorsement to the affected communities, relevant agencies, and within the company. It could be that some government agencies need to approve the plan officially if it relates to their statutory accountabilities. For example, individuals from the local council may have been involved via the Co-ordinating Group during the plan development process, but to gain official approval and adoption, the plan would probably need to be presented to the council as a whole. If communication has been effective throughout, this step should simply be to formalize its adoption.

Agencies such as mines inspectorates, emergency response providers and company management would be targeted at this stage. Group members, and particularly the leader or leaders of the Co-ordinating Group can play a central role as communicators of the plan to expedite its endorsement and adoption.

In cases where the government or official groups may be physically or culturally remote from the area where the APELL process is being developed, gaining endorsement may be slow due to the distances involved, or to scarce resources within the organizations.

Within companies, final endorsement or approval may be required by Head Office. As discussed earlier it is likely that corporate management will have a role in certain decisions and actions in the event of an emergency, and consequently there must also be sign-off on the plan beyond local management. In other cases, the details would be approved locally, but copies of the plans may have to be lodged with the Head Office.

Step 8—Communicate final version of integrated plan to participating groups and ensure that all emergency responders are trained

Once the plan has been endorsed by those groups whose 'sign off' was appropriate or desirable, the details of it need to be communicated to the members of the emergency provider groups so that they are aware of the format of the plan, of their collective and individual responsibilities, and of any training they will require, such as the use of new equipment, new procedures, etc. Operating

Procedures covering aspects of the Plan should be available to all staff who may need them.

- Compile a list of participating groups who will need to know more about the integrated plan.
- Make presentations to these groups to explain the plan, their roles and the type of training they should institute or receive.
- Update procedures manuals.
- Identify those who must be trained; develop and carry out training sessions where necessary. In cases where the local authorities are not equipped to train key people, the mining operation may need to undertake this.
- Ensure notices and posters are displayed in appropriate locations.
- Complete field exercises for hands-on training in monitoring, use of communications, traffic control, evacuation procedures, etc.
- Complete comprehensive workshops, including emergency scenarios, to train leaders in coordination and communication among participants.
- Focus on communication and media training for principal spokespersons in emergency response agencies and within the company. In some cases the media may be one of the response agencies with an important direct role as one of the emergency channels of communication to reach affected people or response providers to trigger plan actions.

These training and presentation sessions may take place as a series of half day seminars. In some instances agencies such as the fire services and environmental agencies may 'cross train' to increase the skill levels of each response team. This has the added advantage of raising awareness of the different issues involved, such as the use of breathing apparatus, contaminant monitoring and containment strategies.

The training should include such issues as:
- roles and responsibilities of responders
- how to use the resources available for a mine related emergency
- procedures for contacting relevant people for information or assistance
- interpretation of UN dangerous goods class, placards and labels
- emergency cards and response guides—how they are structured and how to use them
- location, content and interpretation of documents relating to the contents of a spill
- contact with the media and with other key audiences

Step 9—Establish procedures for periodic testing, review and updating of the plan

The Co-ordinating Group should ensure that the Plan is well tested. Initial testing should take place without involving the public, to uncover deficiencies in coordination among groups and in the training that has taken place so far. Nothing can replace a full scale emergency drill as a means of identifying further areas for improvement. Integrating the drills with other testing procedures on site may be attractive to management, but the potential involvement of different agencies could make this difficult.

- Form a group to prepare a test drill scenario. The group should not include members of the emergency response group.
- Prepare a written scenario that identifies the objectives of the drill, components of the plan to be tested, sequence of events and simulated hazard levels.
- Designate a group of non-participating observers to evaluate the test drill using prepared evaluation checklists.

- Using appropriate local officials, media and other outlets, alert the public and all participants that a test of the plan is scheduled. It is crucial that the public does not confuse the test with the real thing, which could result in panic and a real emergency.

- Conduct the test using the prepared scenario.

- Immediately after the test, the Co-ordinating Group should hold evaluation sessions to consider the results according to the evaluation sheets and the responders' experiences. Interagency cooperation should be a particular focus of this evaluation.

- Assign appropriate participants to correct deficiencies and revise the plan accordingly.

- Prepare a guideline to ensure that the plan is regularly reviewed and updated to keep it current. This should address frequency of reviews in the absence of material changes in the operation or the communities, plus a list of triggers that could affect emergency response and hence should, prima facie, lead to a review of the plan.

Issues to be considered

Multiple scenarios

A different scenario should be tested each time. If there are several scenarios that have very different consequences (e.g. explosion, tailings dam failure, chemical spill during transport to the mine) more frequent testing should be considered until they have all been covered. On a large site which has the potential to affect several different communities, a scenario involving each community should be devised and tested.

Seasonal variation

Weather can raise particular issues or require different responses. Tests should therefore be carried out in different seasons to ensure that plans are as complete as possible. In areas of extreme rainy seasons, for example, access across some rivers may be restricted in times of high flow, so alternative routes may need to be devised. Similarly, in very cold climates, the presence of ice and snow may hamper the plan so that additional resources are needed to allow all areas that may be affected to be accessed.

Varying life span of operations

As well as very different sizes of operations, there are also wide differences in the life span of operations, from a few years to many decades. Emergency response preparedness and the APELL process are obviously applicable whatever the length of life of the operation, but the time taken to step through the process may be deliberately telescoped for a mine with a projected life span of five years, compared with one projected to last 30 years. Consideration also needs to be given to post-closure situations and safety of permanent waste repositories.

Staff turnover

At some operations the mine manager may be replaced regularly, for example, so that continuity in the APELL process may become problematic. This is another reason why regular testing and reviews of plans is a good idea. Not only will it highlight any changes that have arisen affecting the execution of a plan since the last test, but it will also give new managers and emergency providers experience of their roles.

Triggers for review and update of plan

Changed conditions that could require the plan to be reviewed could include such things as an extension to the existing facility, a new area being potentially affected, the development of a new industry in the vicinity of the mine or major new housing or road developments. A substantial change in the capacity or resources of key emergency providers or government agencies may necessitate a review. Other triggers could be:

- a near-miss accident
- a new open pit development
- a new waste rock dump
- a new tailings dam

- a change in process route such as oxide to sulphide
- a new heap leach pad
- other significant changes such as increased mill throughput
- new transport routes or methods

Step 10—Communicate the integrated plan to the general community

Options for involving the community at large, rather than community leaders or representatives, should be pursued at every opportunity throughout the APELL process. The ultimate critical step is to ensure that each member of the community who may be affected knows what the warnings will be and what do to during an emergency, how to get additional information and when to evacuate if necessary. Some awareness campaigns are already reasonably commonplace, for example, to make people familiar with sirens for blast warnings in the vicinity of open pit or open cast mines, in coastal areas for flood warnings, or in buildings and plant to give alerts or order evacuation.

- Prepare a standard emergency response brochure for distribution to all residents in areas that may be affected. This must be appropriate to the level of literacy of the local population—use of symbols and pictures may simplify the response actions, although this may need to be backed up by a face-to-face community education programme. The brochure may need to be in two or more languages for some communities.

- Distribute the brochure by the most appropriate means, such as post, door to door delivery or at community group meetings.

- Prepare a standard media kit which gives emergency contact points in the company, and relevant government and other agencies, as well as providing background information and details on the operation and the emergency response plan.

- Conduct a media briefing session to present the kit and explain what help is needed from the media during an emergency.

- Build other elements of a public awareness campaign such as organizing a pool of speakers available to address local civic groups, schools, etc. special workshops on specific chemicals such as cyanide to educate the public about their benefits and risks. Arrange for media coverage of drill, training activities, etc.

Communications in an emergency

A strategy for handling media contacts during an emergency is a necessary and very important part of the response plan. As noted, the media may have an important role in some cases as one of the emergency channels of communication to reach affected people and response providers. In other cases their role will be informational. It is certainly the case that major accidents involving a mine will inevitably generate rapid and extensive coverage by the news media. As most emergencies, at least initially, are characterized by a lack of information, it is easy to give contradictory messages which lead to the spread of unfounded rumours or misperceptions. These can raise anxiety levels unnecessarily and can be very hard to correct or quash, even after full investigations have been carried out. Regular communication even in the absence of hard data is vital in maintaining public confidence. If no information is available, the spokesperson must undertake to inform the media when more is known.

Proper training and co-ordination is required if contact with the media is to be positive in terms of contributing to the handling of the emergency, providing the community with appropriate information, and limiting damage to the reputation of the company involved. Press interviews and press conferences can be simulated to give personnel practice in handling this aspect of their duties. Building prior contacts with the media staff and involving them at various stages of the APELL process will also help. Supplying information packs about the mine will ensure that they have some facts to use in any report that is made.

Communication is a critical part of the emergency response procedure and one that the mining company must approach professionally. There have been cases of mine representatives appearing on television in the aftermath of an accident who have clearly been ill-equipped to present the facts or to convey the attitude of the company in an appropriate and sensitive manner. The designated company spokesman must have training in handling the media and in communicating key facts and information about the operation as well as being familiar with the emergency response plan before the event. For some emergencies a spokesman will not suffice and the CEO must communicate critical information and messages personally. The use of a spokesperson from the Co-ordinating Group on how an emergency situation is being handled could be considered.

During the planning process it is likely that only local media will be involved, but in the case of a high profile accident, the facility will have to deal with the international media. As part of the planning process, setting up a website for use in emergencies should be considered. Specific details about the mine and its environs can be included, as well as details of the APELL partners and participants. In the event of an emergency, it will be expected that the company—and on some issues the industry association—will provide a continuous stream of updated information, as well as background and explanatory material. A website that is regularly updated is likely to be the most efficient tool for this purpose.

External Affairs, Government Relations and Investor Relations specialists within companies will have their own audiences who will want timely and accurate information on the accident, its impacts, causes and consequences, and response actions taken. They will also be likely to have contact with the media in their roles. Pre-planning, including familiarity with the response plan to be triggered by an accident, will help to ensure that consistent and clear accounts are available to the range of stakeholder audiences.

Chapter 4 of the APELL Handbook and Annex 9 of *APELL for Port Areas* contain more information on the 'Do's and Don'ts' of information communication. Guiding Principles for Crisis Communication are included in Appendix 2 of this Handbook. These are from the European Chemical Industry Council's *Book of Best Practice*, and are highly relevant also to the mining industry. The guiding principles re-iterate aspects to do with preparedness, but also include other principles for dealing with the media in times of crisis. This is an area where much specialist advice is available, and

'As anyone who has been there can tell you, dealing with the media during a period of high stress is an intense experience that is not soon forgotten. However, good crisis management is about more than the care and feeding of reporters. The best media relations in the world will not save a business from damage if it has failed to anticipate problems, to take reasonable steps to head them off before they evolve, and to act quickly and responsibly when they do. Even if the reporters leave with a good feeling toward a company, that doesn't outweigh the potential for damage if it has ignored the informational and psychological needs of its employees, investors and other significant publics.'

David W Guth, Proactive Crisis Communication, Communication World, June/July 1995 this Handbook does not attempt to provide detailed guidance in media relations.

The role of industry associations

Traditionally, industry associations have tended not to speak in the event of an accident, whether it concerns a member company or not. However, this silence is increasingly being seen as unsatisfactory by the member companies, the media, other stakeholders and the associations themselves. Industry associations can usefully play a role separate from the company, which retains primary responsibility for providing accurate and timely information to the various audiences who demand it.

Industry associations should not take over the company role, nor can they pass judgement on the accident's causes, consequences and response measures taken, unless they are involved in a proper process to review those matters. However, they can provide information to help audiences understand the context of the event by, for example, providing descriptions of the mining process used, of the substances which may have been released, or of their impact on human health and the environment. They can point to further sources of information, provide background on the frequency of such accidents and the type and success of remedial actions taken, and provide information on the industry codes to which the company may operate.

As part of being prepared for emergencies, associations should consider drawing up guidelines to define the parameters for their communications before, during and after an accident, provide spokespeople with media training and ensure they have appropriate background information at hand.

Section 5

Mining hazards and risks • Application of APELL in chronic impact situations

The Section gives a description of the hazards and risks which can occur at mining operations. Some of these are common, some less so. Talking openly about hazards and risks is not alarmist, but helps raise awareness amongst communities, government agencies and within companies as well.

Hazards and risks at mining operations

Tailings dam failures

Risks from tailings dams have drawn particular attention recently, following vivid media coverage of several serious accidents that have impacted communities and the environment.

The International Commission on Large Dams (ICOLD) and UNEP have collaborated on a Bulletin entitled *Tailings Dams: Risk of Dangerous Occurrences. Lessons Learned from Practical Experiences.*

> *'Tailings dams can be large and significant engineering works, some of which are amongst the world's major dams...*
>
> *Failure of the retaining dam can release liquefied tailings that can travel for great distances, and because of its greater weight, destroy everything in its path. Water will flow through and around buildings, but liquefied tailings can destroy the structures.'*
>
> ICOLD/UNEP Bulletin Tailings Dams: Risk of Dangerous Occurrences. 2001

The Bulletin contains a compilation of 221 cases of known tailings dam accidents and incidents and provides an overview of causes, lessons learned or remedial actions taken. It notes that during the decade 1979 to 1989 there were 13 significant tailings dam failures. The decade before had at least one failure every year, and the most recent decade, 1989 to 1999, experienced 21 reported failures. In an overall general assessment of the lessons learned, the Bulletin highlights the following causes of failure:

- inadequate management
- lack of control of the hydrological regime
- failure to detect unsatisfactory foundation conditions
- inadequate drainage
- a lack of appreciation of the mechanisms that trigger failures

For big impoundments the failure of a dam and the uncontrolled release of tailings is likely to have serious consequences for public safety, the environment, the owner and the operator.

ICOLD has published a series of guidelines for the design, construction and closure of safe tailings dams. Many other guidelines exist, including management frameworks such as The Mining Association of Canada's *A Guide to the Management of Tailings Facilities* which covers sound management through the life cycle of a tailings facility. The guidelines include emergency response plans (with communications plans) during construction, operation and in the decommissioning closure phases.

Waste dumps

Waste rock dumps constitute another physical risk at mine sites. They are often major structures, consisting of large quantities of overburden or rock containing sub-economic levels of minerals. While some rock may be back-filled in underground mines, or used in the construction of embankments, roads and even tailings dams, high stripping ratios in open pits mean that very large quantities of waste rock may be generated to produce a small quantity of metal or concentrate. Steep terrain and/or high dumps often result in dumps having very long slopes at the angle of repose for the particular material.

Waste dumps have occasionally failed with fatal consequences. The most notable occurrence was at Aberfan in South Wales, where a coal spoil heap engulfed a school in the village, killing 116 children and 28 adults. More recently (2000), part of a 400 m high waste rock dump failed at the Grasberg mine in Indonesia, slipping into a lake and generating a water wave that killed four contractors.

In another case, from Cornwall in the U.K., in the 1990s, a slip occurred in a relatively new china clay waste dump. The slip crossed a road and engulfed a house opposite. Fortunately, the resident was not in the house at the time, but at a different time of day the accident could have been fatal.

Transport to and from the site/loading

The mining industry is a heavy user of road, rail, sea and in some places helicopter transport, usually over long distances, to bring inputs to sites and to remove products, by-products and wastes. In fact much of the industry's business concerns transport and handling of very large quantities of material on and off-site—most of it non-hazardous. Considerable quantities of hazardous materials are also transported in many places by the company itself, in other places by suppliers or specialist transport companies.

Supplies such as cyanide or sulphuric acid may be brought in and transferred between different transport modes in the process—from ship to port to truck to barge to helicopter. Some serious transport accidents have occurred involving cyanide—which has been spilt into rivers in truck accidents, dropped from helicopters, and lost from barges. Wastes and by-products, e.g. mercury, are transported from sites, including over poor roads and through local communities. A transport accident involving mercury had serious consequences for health in the affected community (see Yanacocha Case Study, Section 6). Other transport accidents have occurred in remote areas posing environmental threats, but going unreported.

It is now being realised that the attention given to emergency preparedness for transport accidents may not have been commensurate with their frequency and potential impact and that this is an area requiring greater attention by the mining industry.

Whether or not a mining operation outsources its transport, in the event of an emergency it is the company's reputation that is most likely to be damaged and the company that is likely to be involved in clean-up and remedial activities. It is therefore incumbent on the company to satisfy itself that the transport contractor or supplier has sound emergency procedures in place, along the handling chain. The mine may catalyze joint planning or use a transport accident scenario to test response, communications and community preparedness.

An increasingly common practice is for mining companies to buy reagents only from reliable suppliers who use well qualified, experienced transport contractors. Companies can specify requirements in contracts with suppliers and audit performance.

TransAPELL and *APELL for Port Areas* provide extra guidance on procedures to assist.

Pipeline failure

Pipelines carrying tailings, concentrate, fuel, or chemicals frequently traverse long distances, perhaps between parts of a large mine site or across public land. Facilities such as processing plants or heap leach pads may be several kilometres from the mine itself, and waste disposal facilities such as tailings dams may be further away, particularly in mountainous terrain. This may mean that the mine site is extensive, or consist of several small areas with pipelines, as well as haul roads, between them. Pipeline rupture is a relatively frequent event, rapidly spilling large quantities of materials. This may go undetected for some time and as secondary retention structures are not always provided, released substances enter the environment.

At Browns Creek Mine, in New South Wales, Australia, a leak occurred in a buried return pipe carrying cyanide solution. By the time the leak was detected, the surrounding area was saturated. Close to a boundary fence, a watercourse or a sensitive aquifer, this type of accident has the potential to cause harm to the local community.

Subsidence

Underground mining can result in ground level subsidence over time. Sometimes this can occur without warning, although it is usually a gradual process. Subsidence can occur over relatively large areas, particularly over shallow extensive mining operations such as coal mines under incompetent ground. In other areas subsidence can occur above historic mining sites as structural supports age and deteriorate. Catastrophic failures, while not common, do occur, with buildings, and potentially lives, at risk. Often it is mining communities themselves on top of old mining fields that are exposed to this risk.

Spills of chemicals

Fuels and chemicals used at mining and metallurgical sites are often hazardous in nature, and may be toxic to humans and animals or plants. Virtually all are also in common use in other industries. A relatively small number of chemicals is widely used in the industry in large quantities, and the risks associated with these are well known.

Cyanide has become associated with gold mining although it is used in other industries. The public has an emotional response to its use and any accident involving cyanide can be expected to attract strong media attention. Sound emergency preparedness and communication for operations handling cyanide should be an obvious priority. A voluntary Code for Cyanide Management in the gold industry is being developed, and it contains a strong emergency preparedness component.

For other fuels and chemicals such as sulphuric acid, lime, sodium hypochlorite, etc. the industry needs to be aware of the experience of other industries in relation to storage, handling, risk reduction and emergency response measures directed at chemical spills which can occur in transport, storage or during transfers.

Fires and explosions

Explosives are widely stored and used at mine sites. ANFO is normally mixed at the drill hole. It is shipped as two components: ammonium nitrate (fertilizer) and fuel oil (diesel) both of which are significantly less hazardous than ANFO. Explosives are normally well controlled and stored in safe conditions in magazines which may be operated by the company or a contractor. As noted above, other flammable substances such as fuel (diesel, petrol and kerosene), and sometimes liquefied petroleum gas (LPG) are also transported and stored in large quantities, as are chemicals such as solvents, ammonia, sulphur and relatively smaller quantities of process reagents. Large operations may have acetylene plants for use in the workshops, while small operations will also store and use acetylene. Some mines also have oxygen plants, some use raw sulphur to produce sulphuric acid, while others may use high temperature and pressure together with acid leaches to extract the metals from the ore.

Risk assessments conducted by mining operations are an important tool in highlighting chemicals that may be targeted for substitution to reduce risks, or storage and handling practices which need redesign.

Risks at closed mines

Closed mines can be the source of accidents. Catastrophic releases of contaminated water can occur, such as the release of acidic, metal rich waters from the closed Wheal Jane mine in Cornwall, U.K., in 1991. The event was the result of flooding in the old workings, which were rich in oxidation products from many years of weathering, and the failure of a plug in an adit that released some of the water into a nearby estuary.

In documenting 221 tailings dam accidents, the ICOLD/ UNEP Bulletin records relatively few accidents associated with inactive dams. However, the Bulletin notes that they are not immune to failures, and those which have occurred have usually been as a result of an increase of pool water level resulting in overtopping or slips, and earthquakes.

As mentioned elsewhere in this Handbook, mine stability and safety need to be addressed in the decommissioning and post-closure phases. Contingency planning and emergency response plans need to be tailored to the different situations which will exist. In particular, the mine will no longer have the resources or the personnel to deal with local emergencies. Planning must address the ongoing capacity and roles of the community and government agencies in emergency preparedness for the long term.

Application of APELL in chronic impact situations

As described throughout this Handbook, APELL is a planning process targeted at accident prevention, preparedness and response. However, the same approach and many of the steps can be equally relevant in other situations where community and

Reducing risks from hazardous materials

Guidance is available on best practice in managing hazardous materials to minimize the risk of damage from accidental releases. This can be achieved through:

- knowing which hazardous materials are on site;
- allocating clear responsibility for managing hazardous materials;
- understanding the actual or potential hazard and environmental impacts in transporting, storing, using and disposing of these materials;
- minimizing the use and/ or generation of hazardous materials;
- constructing storage facilities that contain the materials in all foreseeable circumstances;
- disposing of waste materials in a way that eliminates or minimizes environmental impacts;
- implementing physical controls and procedural measures to ensure that no materials escape during normal or abnormal operations;
- having emergency response plans in place to ensure immediate action to minimize the environmental effects should accidental or unplanned releases occur;
- monitoring any discharges and also the environment to detect any escapes of the materials and measure any subsequent impacts; and
- keeping adequate records and reviewing them regularly so future environmental problems are anticipated and avoided.

Hazardous Materials Management, Storage and Disposal. Best Practice Environmental Management in Mining Series, Environment Australia, 1997.

environmental impacts are involved. The approach can be applied in post-accident consultative processes, or in situations where communities may be exposed to long- term chronic impacts from a mining operation. The objective is essentially the same—to use community consultation and planning involving many parties, in pursuit of integrated, effective action which is well understood and supported by all.

Chronic impacts can result from soil and water pollution, due to the migration of contaminants from the site through the physical dispersion of particles by erosion and weathering, or from chemical dispersion such as acid rock drainage or leachate from mine wastes. Long term seepage of fluids from underground storage tanks, tailings dams and pipelines can affect groundwater and surface water. Small-scale 'garimpiero' mining activities can cause long-term effects on local communities, particularly through the use of mercury in gold mining. In areas where rivers are being disturbed, high sediment loads in the river can have longer term effects on fish populations which in turn can have an impact on local communities downstream.

While these are not 'accidents' in the same way that sudden events are, the framework of the APELL process may be utilized to discuss such problems with local residents and authorities, to reach agreement on the most appropriate course of action.

Section 6

Accident case studies

In identifying potential hazards and preparing for all possible emergencies it would seem obvious that a review of past failures would be one useful input to the process. It is not however apparent that all mining companies undertake reviews of past accidents. Scenarios derived from real accidents can and should be applied to contingency planning. The lessons of real situations can help in communicating the risks with communities, as well as in developing effective responses drawn from experience.

The Case Studies in this Section cover quite different situations—tailings failures, transport and subsidence accidents. They bring to life the vivid reality that accidents can have devastating consequences for communities, the environment and companies. They confirm from experience that emergency preparedness is vital and a sound investment for all concerned.

UNEP is grateful to the companies concerned for agreeing to share their experiences through this document. Each case outlines a unique set of events, but collectively they illustrate key issues:

- the importance of risk assessment and proactive risk reduction
- the importance of operational monitoring and checks to give early warning, and the need to act when problems emerge
- the need for contingency plans even when accident probabilities are very low but potential consequences high
- the need for planning and communications to be effective across national or administrative boundaries
- the need for relationships and communications to be in place before any accident
- the need for communities to be aware of the nature of the operations, risks and the properties of the chemicals being used
- particularly, the importance of effective, timely, professional, open communication with the media and others concerned

Case Study 1

Tailings pond failure at the Aznalcóllar mine, Spain

The tailings pond failure in 1998 at the Boliden Apirsa mining operation—a lead, zinc and copper mine located 35 km west of Seville, Spain—captured the attention of the media, industry and public. No deaths or injuries were caused by the spill and no livestock were reported missing. Damage to structures was limited and no major bridges were affected. The nearby Doñana National Park also escaped damage due to timely action to block the waste flow. However, in other ways the immediate and potential long-term effects were severe. The water and tailings affected more than 50 irrigation wells on the river floodplains, and aquatic life in the rivers was depleted. The spill affected farmland used for grazing, agricultural crops and fruit plantations and included important sites for migratory birds.

The mine and concentrator are designed for an annual production of 4.1 M tonnes. The tailings were deposited in a 160 hectare (ha) pond on the banks of the Rio Agrio. Designed and built in 1977/78, the pond contained 15 million m^3 of tailings at the time of the accident. In 1996 it was subject to a full-scale stability study by independent experts and Spanish authorities. No signs of instability were detected at that time. It was subject to regular third-party inspections, the last less than two weeks before the failure. No indication of the problem that was to occur was discovered at that time.

On the night of 24 April 1998, a failure in the marls 14 metres below the dam caused a 600 m section of the dam wall to slip forward by up to 60m. This created a breach in the dam through which tailings and water escaped. Within a few hours, 5.5 million m3 of acidic and metal-rich water had flowed out of the dam. The amount of tailings spilled has been estimated at 1.3–1.9 million tonnes. The spill flooded the riverbanks for a distance of 40 km downstream. In total 4634 hectares of land were affected, of which 2600 ha were covered by tailings. When the water level fell, the depth of deposited tailings ranged from between 4 m near the tailings pond to a few millimetres 40 km downstream. The flood wave was contained in an area downstream by an emergency containment wall built between the river banks. This prevented the contaminated water from reaching the Doñana National Park.

Mining and milling operations were stopped immediately, the breach at the mine was sealed in 36 hours. Spanish authorities banned all use of wells and the affected land. Boliden Apirsa bought the harvest of fruit from the affected area for the season to minimize effects on the farmers and to ensure that no contaminated fruit reached the market place. Boliden Apirsa organized a number of working groups to address various issues including: investigating the causes of the dam failure; environmental impact; clean-up of the spilled tailings; insurance and legal issues; information issues; re-start of the mining operations; and decommissioning of the failed dam. The organization at the mine was not dimensioned to handle this workload and people were brought in from within the Boliden group and external help sought.

A plan for clean-up was presented to the authorities three days after the accident.

Responsibilities were divided by area between the mine and the local authorities, and clean-up had to be complete before the autumn rains. The objective was to return the land to a state where previous land-uses could be continued. The tailings were removed and trucked to the old Aznalcóllar open pit for disposal. While haul roads were used alongside the river, public roads also had to be used, and with hundreds of trucks involved there were five fatal road accidents during the clean-up operation.

After completion, a soil sampling programme was carried out and a second phase of clean-up, in summer 1999, concentrated on areas where soil metal levels exceeded the intervention values. 45 wells were also cleaned.

Approval was granted to use an old open pit as a tailings disposal facility and mine production was re-started. In addition, various works were carried out to decommission the failed dam, including construction of a new channel for the Rio Agrio and building of an impermeable cut-off wall between the dam and the river.

Boliden has highlighted a number of emergency preparedness issues arising from the accident.

- The importance of having relationships in place before an incident not afterwards, in order to build trust between parties and to establish roles and responsibilities, action plans, etc.
- The need for internal and external information cannot be overestimated. Significant resources have to be allocated to dealing with the mass media.
- An information centre was set up in a nearby village but in retrospect the company should have been more proactive in providing information to the local community.
- The benefit of having good baseline data to establish background conditions such as levels of metal concentrations in soils. This would have significantly facilitated the evaluation of the effects of the accident and the clean-up end-point.
- The need to provide employees—unavoidably under great stress at such times—with support as well as information.

The case also highlights that clean-up operations carry their own risks. The large logistical operation that may be required to deal with the aftermath of major accidents may itself necessitate a degree of risk assessment, emergency planning and community communication.

Case Study 2

Mercury spill near San Juan, Choropampa and Magdalena, Peru

Minera Yanacocha SRL operates an open pit gold mine in northern Peru. Mercury is a by-product of gold recovery and is sealed into flasks and then transported away from the mine. On the morning of 2 June 2000, a truck departed from Minera Yanacocha's mine site with a load of 10 empty chlorine cylinders and nine flasks of mercury, each weighing around 200 kg. As the result of a series of events an estimated 151 kg of mercury leaked from one of the flasks and was spread along a 40 km section of highway that passed through three villages, San Juan, Choropampa and Magdalena. The spill allegedly went unnoticed by the driver and was not confirmed until the next day. During that interval, residents of the villages and surrounding areas found and collected quantities of the mercury.

What happened subsequently is open to conjecture. However, there is no doubt that local people directly handled the mercury. In addition some people may have heated mercury in open containers, in poorly ventilated homes, believing it to hold medicinal and religious properties, or in the mistaken expectation of recovering gold. Within a few days many villagers became ill and were soon diagnosed with symptoms of acute mercury intoxication.

In the following days and weeks, between two to three hundred villagers were positively identified as having some level of exposure to mercury, with varying degrees of illness. As in many emergencies, initial responses involved a certain amount of confusion and lack of preparation for such an event.

The consignment of mercury and chlorine cylinders departed the mine and at a point 155 km from the Pan American Highway an empty chlorine cylinder fell from the truck. As the cylinders are too large for one man to handle the driver drove on through the nearby settlements of San Juan, Choropampa, and Magdalena. The next day a supervisor came to meet the driver and they retraced the route to the chlorine gas cylinder. People were at that stage collecting mercury in the street but this was not remarked upon, and the driver appeared unaware that anything had been amiss with his mercury load. On the same day the duty manager at the mine received a call from a friend living in Choropampa stating that there appeared to be mercury on the streets. Two hours later personnel from the mine arrived to investigate, and found a child playing with what appeared to be mercury. A team was therefore sent to try to recover the remaining spilt mercury. Meanwhile, the truck carrying the mercury had been left unattended in Magdalena, the flasks were in disarray and around three-quarters of a flask was estimated to be missing.

Over an extended period loudspeakers were used, meetings held and advertisements put in local papers warning people that the mercury was poisonous and to return it to the medical posts in the villages, but efforts were frustrated because villagers refused to return the mercury. Over the next few days repeated efforts were made to recover the mercury and several people reported to health centres with contact dermatitis caused by handling it. At this stage (six days after the spill) the health authorities were unaware of the risks posed by the inhalation of mercury fumes, believing that skin contact and ingestion posed the only threats. The number of people falling ill increased over the next four weeks, and Minera Yanacocha began to buy back the mercury. It is estimated that approximately 45 per cent of the mercury was returned or recovered through road clean-up operations and that approximately 15 per cent was lost to the air through vaporization. The rest remains unaccounted for. A total of 511 people had been treated for some form of mercury poisoning by 23 July, 134 of them in hospital.

The report produced by the Compliance Advisor/Ombudsman of the International Finance Corporation notes several points that hampered response to the situation

- Lack of an emergency response plan dealing with spills off the mine property.
- Confusion regarding 'ownership' or responsibility for the incident (transport company versus the mine).
- In many cases a lack of cooperation from the local population.
- Poor communication between the company and local authorities.
- Remote location of the spill delaying delivery of clean-up and analytical equipment for clean-up.
- Confusion over how much mercury was actually lost.
- Distribution of mercury over a wide area.

Follow-up action

Since the spill, the Ministry of Energy and Mines has published a resolution ordering mining operators to submit contingency plans and operating manuals for hazardous or toxic substances. An initial list of substances that could present some level of risk or concern to health or the environment for which contingency plans should be developed includes cyanide, mercury, sulphuric acid, fuels and lubricants, lime, sodium hydroxide and hydrogen peroxide.

The company's follow-up action includes:

- A road transport hazardous material supervisor is being hired who will supervise transport at loading and unloading points as well as on road checks.
- A new contract has been signed between Yanacocha and the security contractor for road traffic control between the coast and Cajamarca. A checkpoint at the beginning of the journey is being set up at which all trucks and drivers will be inspected to ensure their good mechanical and physical condition.
- Transport of hazardous material will be done in convoys and only during daylight.
- Training sessions will be held for local authorities and communities to explain more about operational activities and hazardous material being used at site, including the emergency plans and how they can help to reduce risk if needed.

Case Study 3

Marcopper tailings spill, Marinduque Island, Philippines

On 24 March 1996, 1.6 million m^3 of tailings flowed through an old drainage tunnel from a closed pit where they had been stored at the Marcopper mine. The tunnel had been sealed prior to tailings disposal, but the plug failed. The failure may have been aided by a small earthquake six days beforehand.

The release caused major disruption to the local inhabitants. Although people did not draw water from the river to drink, it was used for washing clothes, as well as irrigation and as a source of water for livestock. The river is also used as a road for most of the year, and the spill meant that vehicles could not cross the river, cutting off some communities upstream from the local town where they sold their goods in the market. Crops on the riverbank and fish traps in the river were inundated or destroyed. As the mine was closed immediately, workers were laid off and benefits to the local economy in the form of wages, services, supplies and local taxes dried up.

At the time of the accident, the mine was owned 39.9 per cent by Placer Dome, 49 per cent by the Philippines government and 11 per cent by the public of the Philippines. Placer Dome, although not the majority shareholder, took responsibility for the remedial work and for compensating those affected, and also for creating a sustainable development programme that will provide a positive legacy and long term benefits to the community.

Company staff have summarized lessons learned from the incident:

- Gathering accurate information is time consuming and expensive, but essential.
- Short term technical fixes are often very different from long term solutions faced by local communities.
- In developing long term solutions it is difficult to please everyone, but this should not preclude action.
- When it comes to community outrage, 'perception is reality.'
- Communication should take place in non-technical language.
- Involving the community in decision making from the beginning is essential—things may seem to take longer but community decision making will actually speed up success.

'Throughout this experience, we have certainly learned how important it is to involve people in decisions that affect them. Engineering experts have a tendency to make plans then advise people what they will do—especially in an emergency situation. However, in our experience, people often object to these plans for the simple reason that they are not 'their' plans. That is not to say we can always do what the people want—there may be technical reasons that prevent it. But some people still need to be consulted as part of the decision-making process.

In the end, mining is not about minerals, it's about people. Success will only come to those who demonstrate commitment to the communities in which they operate and are capable of building trusting relationships with them.'

John Loney and Christopher Sheldon,
Placer Pacific

- Time and energy spent developing community goodwill on a daily basis can be a valuable asset in times of crisis.
- When there is a breach in the relationship such as that caused by an accident of this type, it takes time to rebuild credibility and trust.

Effective communication was a key issue. There is a great demand for information at many different levels in the aftermath of an incident of this nature, and while most mining companies are good at producing quality, detailed technical reports ideal for government departments, these are often of little use to people in local communities. In this instance, Placer created an illustrated story book with dialogue in the local language as well as English which has been effective in helping people understand events and subsequent activity.

Case Study 4
Failure of tailings dams at Stava, northern Italy

At a fluorite mine near the village of Stava, in the mountains of northern Italy, two tailings dams were built across a valley, one upstream of the other. A stream running down the valley was channelled through a concrete pipe laid in the bed prior to dam construction. When the first dam reached a height of 16 m, construction of the second dam began upstream of the first impoundment. When this second dam reached a height of 29 m it suffered a rotational slip and breached. It is thought that the water pipe and decant structure failed allowing the stream to discharge into the body of the lower part of the upper dam. The resulting rise of pore pressure caused the rotational slip and the failure. The released tailings then caused the failure of the lower dam and the combined contents of the two impoundments flowed at speeds up to 60 km/hour sweeping away the village of Stava with its several hotels and engulfing part of the small town of Tesero, 7 to 8 km downstream. 269 people were killed.

An emergency response planning procedure including a sound risk assessment process could have identified unacceptably high risks for the community, the authorities and the mine. In hindsight, some leading engineers believe that the site was fundamentally unsuitable and the risk-weighted consequences of a failure of the dam too great for it to have been built. It is not clear that effective emergency response plans could have been developed, given that the speed with which the accident occurred left no time to warn the community. However, an APELL type process would at least have raised the profile of the risk. The option of relocating people living below high risk tailings dams has been used when companies have had the imagination and wisdom to act to prevent the unthinkable.

Case Study 5

Cyanide spill at Baia Mare, Romania

On 30 January 2000, a breach occurred in the tailings dam at the Aurul SA gold and silver mine in Baia Mare, Romania. Some 100,000 m^3 of liquid and suspended waste were spilled, containing an estimated 50–100 tonnes of cyanide as well as some heavy metals, particularly copper, into the river system. The contamination travelled via tributaries into the river Somes, Tisza and finally into the Danube, affecting Romania, Hungary and the Federal Republic of Yugoslavia.

The mine was set up to reprocess old tailings and commenced operations in 1999.

The breach in the dam was caused by a combination of inherent design deficiencies in the process, unforeseen operating conditions and bad weather. The dam was being constructed using coarse tailings from the operation, a method which requires a safe level of freeboard to be maintained between dam height and pond water level.

In the case of the new Aurul pond at Baia Mare, the flows of solids and waters were out of balance with the increase of the storage capacity of the pond, as the process of dam construction could not keep up with the rise in the reservoir water level. The climatic conditions of the winter season aggravated the situation and led to an uncontrolled rise of pond level resulting in an overflow of the dam.

The company repaired the breach using borrow material from nearby, and added sodium hypochlorite to the overflow (and to the area flooded by the spill). A large volume of heavily contaminated effluent escaped before the breach could be closed.

The pollution had the potential to severely impact biodiversity, river ecosystems, drinking water supply and socio-economic conditions of the local population. Acute effects, typical for cyanide, occurred for long stretches of the river system down to the confluence of the Tisza with the Danube: phyto- and zooplankton were down to zero when the cyanide plume passed and fish were killed in the plume or immediately after. These effects were not long lasting.

The report of the UNEP/OCHA Assessment Mission on the accident concluded that the company took reasonable steps to respond to the emergency. Also, that the early warning system established under the Danube River Protection Convention responded adequately to alert neighbouring countries. Timely information exchange and measures taken by the Romanian, Hungarian and Yugoslavian authorities, including a temporary closure of the Tisza lake dam, mitigated and reduced the risk and impact of the spill. Villages close to the accident site were provided with alternative water sources.

However, the report noted that there appeared to be no monitoring system to detect the onset of dangerous situations. Furthermore, formal emergency preparedness and response procedures by the company and local authorities were rudimentary considering the large quantities of hazardous materials (cyanide, hypochlorite) being used close to human populations and the river system. The report noted delays in getting information to the population in the vicinity of the plant at the earliest desirable stage and considered the establishment of a good operational and prompt early warning system essential. The accident also exposed the fact that there was little public awareness in the local population of the environmental and safety risks which can occur at mining operations.

An APELL process may have ensured that the risk of the accident was foreseen, and should have ensured that the mining company had open lines of communication not just with officials, but with the local community. The mining company may also have been better prepared to deal with the international media attention that the incident attracted.

Case Study 6

Subsidence of active mine workings, the Lassing talc mine, Austria

This case study concerns a subsidence accident at a talc mine owned by Rio Tinto, through its wholly owned subsidiary companies Luzenac and Naintsch Mineralwerke, and located in the Province of Steymark, Austria. The accident caused fatalities within the mine boundary and had far-reaching physical and emotional effects on the local community. The mine has not operated since the disaster and will be permanently closed in 2001.

This underground mine had been worked since 1901 and produced some 25,000 tonnes of talc per year. The mine and its associated mill, which continues to operate, are situated in a small valley midway between Vienna and Salzburg. Mining was carried out by the underhand cut and fill method. At about 10 a.m. on Friday 17 July 1998 a miner, Georg Hainzl, became trapped underground in a rest room on an upper level following an inrush of water and mud. A crater began to appear on the surface and houses located in close proximity to it began to tilt and move. The entire mining workforce returned to site to assist in the rescue. Company officials came from Graz, the headquarters of Naintsch, and officers from the Provincial and Federal Mining Authorities arrived from Vienna and Leoben. By the mid afternoon, the site was overwhelmed by the media, representatives of various authorities, fire brigade officers, local community members, police, friends and family of the trapped miner and the rescue miners, and general onlookers. In all, some 700 people were at, or around, the mine site. Local and Graz management together with the officers of the mining authority and the rescue team spent much of the afternoon and early evening underground, planning and effecting one of the rescue plans.

Suddenly, at about 9:30 p.m., there was a loud noise and houses on the surface began to slide into the crater. The crater rapidly increased in size and filled with water. Those at the pithead felt a violent rush of air expelled from the shaft. At that point it was realized that a catastrophe had struck and that the nine miners and one technical expert, who were still underground as part of the rescue effort, were in terrible trouble.

After nine days of frenetic activity, Georg Hainzl was rescued via a drill hole from the surface. This raised hopes that the others may also have survived and rescue attempts therefore continued for a further three weeks. They were halted on 14 August 1998. Various plans to re-enter the mine to seek and recover the bodies of the rescue team were then worked on. In April 2000 these plans were finally put aside for safety reasons and planning for final closure of the mine was commenced.

To put this tragedy into perspective, the nine Lassing miners comprised almost the entire mining workforce. Most lived in and around the Lassing village and had relatives and family living within 5 km of the site. Some relatives, including brothers, fathers and sisters also worked in the mill. One house was destroyed and two were so badly damaged that they had to be demolished. Some 12 families had to be relocated. The main local road and a local stream were cut and had to be diverted. This accident therefore deeply impacted a very small, tightly knit community.

Investigation of the accident continued throughout 1999 and court action against the mine manager and several members of the Mining Authority personnel took place during 2000. There was therefore almost continual media coverage and exposure of the accident for more than two years after its occurrence.

A tragedy of the scale that occurred at Lassing was a significant event for the country. Apart from the relatives, families, employees and the company management, who were directly affected, other parties or groups became involved. These included Rio Tinto and Luzenac, the media, politicians, local community and a wide range of technical experts.

The government and the company provided counselling and caring services, which included group therapies and one-on-one sessions for the relatives, group discussions with employees and psychiatric counselling for the mine manager and Georg Hainzl.

As with most newsworthy incidents, the media (press, radio and television) played a prominent part in, and had a substantial influence on, how things developed. The situation at Lassing possibly experienced greater exposure because of the large crater that formed (100 m diameter and 40 m deep) and which, for reasons of investigation and approval by the authorities, was only filled in more than two years after the event. This constant reminder ensured that Lassing would never be far from public attention. It was noticeable that, until and throughout the trial, pictures of the crater usually accompanied news reports.

The media behaved in three different ways. Initially, due to lack of quality information, as a vehicle to establish what happened and to promote the recovery of the bodies. Then, as a supporter of the relatives when it appeared that investigation/explanation and recovery were progressing slowly. Finally, as a voice for the defence at the trial, where it appeared that the prosecution was disallowing presentation of some evidence. Their change in attitude as time progressed appeared to result from better management of the information flow. The company eventually developed a strategy to supply as much information, and in as simple a manner, as possible, whereas, for some time after the accident, there was no concerted, planned effort to keep the media in the picture.

Politicians and government departments at the local, Provincial and Federal levels also became deeply involved. The main problem arising from the political involvement at Lassing was one of lack of understanding. The mechanism which lead to the catastrophic inundation at Lassing was very complex and a complete explanation for the tragedy has yet to be found, even after 18-months of determined investigation and a six-month trial. Yet, the politicians understood that the families of the deceased miners wanted to recover their bodies and they promised that this would happen on the false assumption that this was only a cost issue. However, it was clear, probably from around November 1998, that recovery of the bodies posed too great a safety risk and was not practicable. It was not until April 2000 that a statement emerged from the Ministry that it would not be possible to recover the bodies. By then the families and relatives were well aware that the bodies would not be recovered.

Lassing is a small community, comprising about 500 families, which was thrust into the limelight by the disaster. Because the community knew or understood very little about the situation, much misinformation circulated within it. Spurious accusations of waste material being dumped in the mine, illegal mining, management arrogance, major settlement of houses over the last 10 years, noise from blasting and so on, were all raised.

Initially, it was felt by those involved in the investigation that the technical issues were too complicated for the community and general public to understand. Rather than trust in their ability to comprehend the problems they were excluded from the process. This problem was realized early in 1999 and the Mayor of Lassing was invited to join the weekly progress meetings held between the company, the Mining Authority and the relevant Government department representatives. The Mayor thus came to understand that recovery would be difficult and dangerous. It was also noticeable from that moment that the relatives of the deceased no longer directed their frustration and anger at the company as they had done immediately after the accident

The company believes that there are some clear lessons to be learned from this accident.

- Information is best when it is managed and when it comes from the company, because it is the company that has the most accurate and up-to-date information.
- The immediate appointment of a senior credible spokesperson is crucial.
- Inclusion of affected stakeholders at an early stage will help dispel rumour and ill-feeling toward the company.
- Have an emergency response plan in place.
- The plan would avoid the chaos on site during the hours as a crisis develops.
- The plan should include specific strategies and methods for dealing with the media, community groups and government bodies.
- Strategies need to take into account particular cultural contexts as well as the specific nature of the operation.

Section 7

Appendix 1
Components of an emergency response plan

The following suggests items which might usefully be documented in a written emergency plan. Not everything listed here would necessarily be included or, alternatively, different things may be required in some cases. The order is not very important. The list has been compiled from a number of sources and is intended as an aid; it should be applied flexibly.

Purpose/objectives/scope

Objectives of the plan
When the plan is to be used and by whom
Defining an emergency
Emergency scenarios covered
Other elements included in the document
Date of plan/ frequency of updates

Emergency scenarios and risks

Emergency scenarios separately identified/ outlined
Population and residential centres at risk
Nature of environments/ wildlife populations at risk (Baseline monitoring data recorded elsewhere)
Maps of risk areas/ modelled airshed, watershed impacts
Quantity and location of hazardous substances
Properties of each hazardous substance (MSDS sheets and UN designation)

Mine Emergency Co-ordination Centre

Designated person in charge/alternates
Location of Mine Emergency Co-ordination Centre/alternate location
Role of Centre
Communications systems/ equipment to reach other emergency/response providers
List of functions of key people (on-site, off-site)
List of telephone numbers (office, home, mobile) for key people/alternates
Centre to hold key documents, e.g.

- shutdown procedures for operations
- locations of hazardous material storage areas/ emergency and safety equipment
- maps of communities and environmental maps
- information on location of other communications equipment, including portable sets
- information on emergency power
- contacts for other utilities
- operating manuals
- MSDS sheets
- list of personnel with alternate skills for use in emergencies
- type, location of alarm systems
- accident report forms
- accident status board and log book
- copies of emergency plan, media and communications plan, specific action plans
- notification lists, staff lists, contact lists, with regular and emergency telephone/pager numbers, etc.

Media and Crisis Communication Centre

(see Appendix 2 for more guidance)
Person in charge
Location (co-located with Emergency Co-ordination Centre, separable, remote)
Role of Centre
Inter-relationships/ links with Emergency Co-ordination Centre
Roles of individuals in communication team
Lists of contact details for media, NGOs, politicians and officials, investor, other individuals from key audiences
Media briefing facilities
Communications equipment
Procedure for internal communication
Procedure for external communication
Procedure for notifying families of those injured
Procedure for media contacts
Pre-established website
Designated trained spokesperson(s)
Custodian of:

- crisis communication plan (media, etc.)
- communication guidelines
- key messages
- background information on operation and emergency response plan
- log book of contacts/ statements made

Emergency notification procedures and communication systems

Information to be given when assessing the incident

- proformas

Criteria for determining levels of alert
Contact details for primary alerts (immediate emergency responders)
Contact details for secondary alerts (need to know, standby, etc.)
Emergency notification flow chart

- organization to be contacted
- by whom
- method of communication
- sequence

Alarm systems
Individual's names and telephone numbers, with alternates, for, e.g.:

- plant manager
- local officials and response agencies
- neighbouring industrial facilities
- community leaders
- nearby residents
- media

Communication equipment (radios, mobile phones, etc.)
Procedure for recording actions taken and communications made.

Emergency equipment and resources

Includes on site and external resources as identified in planning or specified under mutual aid agreements
Location of isolation valves
Special procedures, e.g. specialist fire-fighting, chemical neutralization
Equipment for combating pollution, e.g. booms, skimmers, pumps, absorbents, dispersants
Internal and external emergency medical support

- hospitals
- clinics
- ambulances
- medical supplies
- personnel with medical or first aid training

Earth moving equipment, specialized equipment, boats and landing craft, if necessary
Helicopters—availability, landing sites and refuelling capability
Fire fighting equipment
Toxicity testing facilities—gas and water
Wind direction/speed indicators
Local or regional weather forecasting service
Self contained breathing apparatus
Personal protective equipment
Other capabilities according to risks identified

- in community, government agencies, on site, at other facilities

Containment capabilities and waste disposal arrangements

Emergency scenarios and emergency response procedures

For each emergency scenario (toxic gas release, earthquake damage to tailings dam, chemical spill during transport, etc.) specific and particular information and guidance should be given, e.g.:

Triggering the plan
Command structures
Roles and responsibilities, e.g. shift manager, environmental manager, security officer, external affairs manager
Transfer from one planning level to another
Response actions
Equipment
Notification procedures
Communications procedures
Alarm systems
Evacuation procedures/assembly points
Media procedures
Medical procedures
Assess, monitor and record progress of the accident e.g. time, duration, quantity, location of material released to the environment
Procedures for operational shutdown if necessary
Record of actions taken to respond
Deactivation of the plan

Clean-up, remediation, procedure for returning to normal operations

Options for clean-up plan (including risk assessment on alternatives)
Specified authority to order re-start (site manager, other)
Procedures, key people to inform of re-start
Preliminary investigation including photography, securing of evidence, damage evaluation
Post-emergency evaluation of effectiveness of plan and response
Ongoing communications plan in relation to progress of clean-up/remediation

Training and drills

Should focus on the following:
Possible scenarios for different emergencies
Evacuation of non-essential personnel at mine site
Evacuation of personnel from surrounding area (procedures, shelters, assembly points)
Knowledge of chemicals (properties, toxicity, etc.)
Procedures for reporting emergencies
Operation and knowledge of alarm systems
Operation and knowledge of communications systems
Location and use of fire-fighting equipment
Location and use of protective equipment (respirators, air cylinders, protective clothing, etc.)
Decontamination procedures for protective clothing and equipment
Awareness of clean-up and remediation actions to be taken
Alert media and community before major tests
Document test
Evaluate and revise emergency plans, communications plans
Specify frequency of testing, triggers for new drills

Appendix 2
Guidelines for Crisis Communications from Responsible Care Guidelines of the European Chemical Industry Council (CEFIC)

Guiding principles of crisis communications

- Successful crisis communications start with open communications with all target audiences.
- Public acceptance depends on corporate behaviour before, during and after the crisis, not purely on the nature of the crisis.
- The only consistent element in all crises is the media attention.
- Prepare a 'worst case' scenario.
- Ignoring an issue is inviting a crisis. Preparation is the only way to handle the unpredictable.
- Take control of the situation and be the main source of information.
- The manner in which the first 24 hours of a crisis are handled is the most crucial.
- Don't get involved in speculations on reasons and responsibilities.
- Show concern to all groups involved.

Before the crisis—preparedness planning

There are three steps of preparedness planning:

1. A risk assessment of a company's vulnerable points in terms of people, products, processes, practices and policies.

2. A crisis plan and a crisis manual to help managers in the early hours of a crisis are key elements in gathering the right resources and information and taking the correct crucial first steps. A crisis plan answers the who, what, where and when of crisis communication:
   - who will be in the crisis team?
   - who are the likely audiences involved?
   - what are the basic corporate messages to be delivered?
   - what are the basic systems needed for rapid and thorough communications?
   - when will you communicate?
   - where will the crisis team be located? Corporate, national or local?

(An example of a crisis plan is shown below).

3. Crisis training, where managers who will form the crisis team, experience a series of crisis simulations to hone their skills, test the crisis plan and assess their abilities to develop plans and messages when 'under fire'.

During the crisis—action planning

The elements of a crisis are always the same: surprise, a loss of control, a lack of information and a sense of siege.

When time allows, the following communication steps should be taken into account:

1. Define the problem and set the objectives depending on the nature of the crisis.

2. Formulate the communications strategy. The most effective messages will be those that are in tune with the actual attitudes and perceptions of the audiences. A fundamental motto of message development is 'concern'.

3. Manage the communication process. After a crisis has become public, the company's most important audiences—the local community, employees, customers, suppliers, shareholders, the media and many more—will have many questions. Communication is the process by which a company can keep these audiences informed and therefore, hopefully, on its side until a problem is solved.

Inevitably, the media are the least controllable of all communications audiences and channels in a crisis. A company must be prepared to give honest answers to their basic questions:
- what happened?
- why?
- what action is the company taking over it?

A leading principle in all communications during a crisis is to centralize all communications and to have one single spokesperson to ensure:
- consistency of message
- the audience focus on one person who becomes trusted
- hat others are freed to concentrate on solving the crisis.

Other important principles are:
- Wherever possible, inform employees first. Each employee is an important vehicle for information to the local community. So they need to be informed of the facts.
- Top management should be visible to the public at an early stage, to focus on principal aspects of the event, to express sympathy or concern with those affected, and to assume responsibility.
- Informing management and sending press materials and handling instructions to all other relevant company locations.
- Not to speculate. Describe the facts presently known. Make more information known as soon as it becomes available.
- Publicizing what has been done immediately. Others involved—fire service, police—will do so anyway.

After the crisis—evaluation planning
- Post crisis communication is necessary to optimize resumption of normality.
- Learn from the crisis: evaluate and update the crisis communications plan.
- Communicate to all relevant audiences—internal and external—the follow up actions that have been taken and details of the learning experience.
- Continue developing crisis communication plans.

Contents of a proposed crisis communications plan
1. **Message from the Board** on the importance of a Crisis Communications Plan.

2. **Overview of possible crises or calamities.**

3. **Key contacts/accountables and their responsibilities:**
- central crisis coordinator
- facility crisis coordinator
- safety employees
- central PR coordinator
- facility PR coordinator
- others

4. **Reporting procedures:**
- list of internal and external contacts who have to be informed
- list of key telephone, telex and fax numbers
- communications overview (also in case some employees cannot be reached)

5. **Crisis centre:**
- information on location of central crisis centre
- information on location of facility crisis centre
- list of facility crisis team
- communications strategy with external institutions
- communications facilities
- central media centre
- facility media centre and facilities for media

6. Procedure for internal communication
- family victims
- works council
- personnel
- examples of internal messages

7. Procedure for external communication
- surrounding people
- official institutions
- suppliers
- distributors
- customers
- branch organizations

8. Procedure for media contacts
- rules for facility entrance for media
- rules for corporate and local spokesperson
- media list
- list of do's and don'ts
- guidelines for ensuring accurate and official information
- guidelines for giving information by telephone
- guidelines for interviews
- examples of key messages—public statement, statement to employees, statement to vendors, customers or other facilities
- checklist of possible questions and hints on how to answer them

9. Follow-up
- relatives of victim(s)
- note of thanks to personnel
- letter of thanks to external helping organizations
- letter of thanks to vendors, distributors, customers
- evaluation of the crisis handling

Appendix 3
List of publications related to disaster prevention and preparedness

These documents have been published over the last decade and have become a valuable source of information for accident prevention, risk assessment and emergency response planning.

Order Information:
All UNEP DTIE publications are available from:
SMI (Distribution Services) Ltd.
P.O. Box 119, Stevenage
Hertfordshire SG1 4TP, England
Tel: + 44 (1438) 748 111
Fax: +44 (1438) 748 844
E-mail: enquire@smibooks.com

UNEP Division of Technology, Industry and Economics—Publications:

APELL Handbook
The APELL Handbook provides the basic concepts for the development of emergency response plans at the local level, with an emphasis on community awareness of potential dangers and preparedness for all contingencies.

Storage of Hazardous Materials
This technical report introduces practical guidelines for safe storage of hazardous materials, including information on key responsibilities, legal frameworks, product evaluation, warehouse siting, management, and fire and environment protection.

Hazard Identification and Evaluation in a Local Community.
This technical report describes the hazard analysis method and gives concrete examples of how to implement it. The report also contains several valuable annexes that provide information to enable local communities to identify and evaluate hazards.

APELL for Port Areas
This Handbook sets out the procedure to enable decision-makers and technical personnel to improve community awareness of activities involving hazardous substances in port areas and to improve or create coordinated emergency response plans. (prepared with the International Maritime Organisation—IMO).

TransAPELL
This technical report applies the APELL process to the transport of dangerous goods. Includes information on conducting workshops and developing community response plans, as well as case studies on TransAPELL implementation.

APELL Worldwide
This review contains case studies exploring the adoption and adaptation of the APELL process in 12 countries around the world.

Management of Industrial Accident Prevention and Preparedness
A training resource kit for use in universities and colleges.

Related joint publications with international organizations

These publications are available from the underlined organization.

Tailings Dams: Risks of Dangerous Occurrences. Lessons Learnt from Practical Experiences. ICOLD/UNEP Bulletin, 2001

Proceedings of the International Workshop on Managing the Risks of Tailings Disposal, Stockholm, 1997. (ICME, SIDA, UNEP)

Proceedings of the Workshop on Risk Management and Contingency Planning in the Management of Mine Tailings, Buenos Aires, 1998. (ICME, UNEP)

Guiding Principles for Chemical Accident Prevention, Preparedness and Response
Guidance for Public Authorities, Industry, Labour and Others for the Establishment of Programmes and Policies related to Prevention of, Preparedness for, and Response to Accidents Involving Hazardous Substances (OECD, 1992). (Draft) revision: http://www.oecd.org/ehs/ehsmono/ACGUCON.HTM

Manual for the Classification and Prioritisation of Risks due to Major Accidents in Process and Related Industries (IAEA, UNEP, UNIDO, WHO, 1996).

Guidelines for Integrated Risk Assessment and Management in Large Industrial Areas, (IAEA, UNEP, UNIDO, WHO, 1998).

Health Aspects of Chemical Accidents (IPCS, OECD, UNEP, WHO).

International Directory of Emergency Response Centres (in cooperation with OECD, 2nd edition in preparation).

List of Acronyms

ICOLD	International Commission on Large Dams
ICME	International Council on Metals and the Environment
SIDA	Swedish International Development Co-operation Agency
OECD	Organisation for Economic Co-operation and Development
IAEA	International Atomic Energy Agency
UNIDO	United Nations Industrial Development Organisation
WHO	World Health Organisation
IPCS	International Programme on Chemical Safety
UNEP	United Nations Environment Programme

Appendix 4
List of websites related to disaster prevention and preparedness

UNEP DTIE APELL Homepage:
http://www.uneptie.org/apell/home.html

Risk assessment software:

CAMEO®

CAMEO ® is a system of software applications used widely to plan for and respond to chemical emergencies. It is one of the tools developed by US EPA's Chemical Emergency Preparedness and Prevention Office (CEPPO) and the National Oceanic and Atmospheric Administration Office of Response and Restoration (NOAA) to assist front-line chemical emergency planners and responders. They can use CAMEO to access, store, and evaluate information critical for developing emergency plans.

The CAMEO system integrates a chemical database and a method to manage the data, an air dispersion model, and a mapping capability. All modules work interactively to share and display critical information in a timely fashion. The CAMEO system is available in Macintosh and Windows formats.

To get the complete CAMEO package (CAMEO, ALOHA, and MARPLOT):
http://www.epa.gov/ceppo/cameo/index.htm

Emergency Response Guidebook 2000
The Office for Hazardous Materials Safety
http://hazmat.dot.gov/gydebook.htm

For first responders in the initial phase of dangerous goods/ hazardous materials incidents the *Emergency Response Guidebook* (ERG2000) was developed jointly by the US Department of Transportation, Transport Canada, and the Secretariat of Communications and Transportation of Mexico (SCT) for use by fire-fighters, police, and other emergency services personnel who may be the first to arrive at the scene of a transportation incident involving a hazardous material. It is primarily a guide to aid first responders in (1) quickly identifying the specific or generic classification of the material(s) involved in the incident, and (2) protecting themselves and the general public during this initial response phase of the incident. The ERG is updated every three years to accommodate new products and technology.

Disaster management information on the world wide web

Valuable information of all aspects of disaster management can be obtained from the Internet. The listing below is a small selection of useful websites.

Mining accidents

Mineral Resources Forum (MRF)—Environment
General and timely accident information as it occurs
(http://www.mineralresourcesforum.org)

UNEP/OCHA
Cyanide Spill at Baia Mare: Assessment Mission Report (http://mineralresourcesforum.unep.ch/BaiaMare/index.htm)

UNEP/OCHA
Mining waste spill from the Baia Borsa processing complex in Romania: Assessment Mission to Hungary and Romania UNDAC Mission Report (http://mineralresourcesforum.unep.ch/BaiaMare/docs/BaiaBorsa/baiabrosa-final.pdf)

The International Finance Corporation (IFC)/World Bank Group
Independent Commission Report on the Mercury Spill in the Province of Cajamarca, Peru (http://www.ifc.org/cao/prelease/prelease.html)

International Commission for the Protection of the Danube River (ICPDR)
Regional Inventory of Potential Accidental Risk Spots in the Tisa Catchment Area of Romania, Hungary, Ukraine and Slovakia (http://www.tisaforum.org.yu/defyu/engl/program-icpdr1.htm)

Safe tailings design

The International Commission on Large Dams (ICOLD)
Promotes progress in the establishment of design, construction, operation and maintenance of large dams (http://genepi.louis-jean.com/cigb/)

WISE uranium project
Safety of Tailings Dams, Current Issues—Tailings Dam Safety, and Properties of Tailings Dams (http://www.antenna.nl/wise/uranium/mdas.html)

Industrial accidents

OECD/chemical accidents
Prevention of, preparedness for and response to chemical accidents (http://www.oecd.org/ehs/accident.htm)

UN/ECE
Convention on the transboundary effects of industrial accidents (http://www.unece.org/env/teia/)

European Commission
DG XI: Chemical Accident Prevention, Preparedness and Response: Seveso Directive (http://europa.eu.int/comm/environment/seveso/index.htm)

Hazmat Central
Hazardous material: Managing the incident (http://www.hazmatcentral.com)

Chemicals

UNEP/chemicals
Chemical information resources, including an Internet guide (http://www.chem.unep.ch/irptc/)

IPCS
Prevention and management of chemical emergencies (http://www.who.int/pcs/)

IOMC
sound management of chemicals, of particular interest: a list of meetings and an Internet guide to the activities and programmes of participating organizations (http://www.who.int/iomc/)

ILO
International Safety and Health Information Centre (CIS), contribution of ILO to IPCS (http://www.ilo.org/public/english/protection/safework/cis/index.htm)

EPA
Chemical Emergency Preparedness and Prevention Office (www.epa.gov/swercepp)

General Chemistry-related Information on the Internet (http://www.faqs.org/faqs/sci/chem-faq)

Fires

Global fire monitoring centre
near-real time information on forest fires
(http://www.ruf.uni-freiburg.de/fireglobe/)

UNEP/GRID
global forest and other wildfires status reports, maps, environmental datasets
(http://www.grid.unep.ch/fires/)

Fire and Safety Directory
Useful fire safety information
(http://www.firesafe.com)

Earthquakes

Global earthquake response center
Latest earthquake information, detection, reporting and news (http://www.earthquake.org/)

Earthquake Information
Recent Global Events—Near-real-time Earthquake Bulletin is provided by the National Earthquake Information Service (NEIS) of the U. S. Geological Survey (http://civeng.carleton.ca/cgi-bin2/quakes or
http://gldss7.cr.usgs.gov/neis/qed/qed.html)

Earthquake hazards and preparedness
General information, links, reducing hazards, research (http://quake.wr.usgs.gov/) and
(http://quake.wr.usgs.gov/prepare)

RADIUS
ISDR initiative to reduce urban seismic risk in the world (http://www.geohaz.org/radius)

Floods

Floodplain management association
(http://www.floodplain.org)

Tornadoes/Hurricanes

Tornado project online!
(http://www.tornadoproject.com)

Significant tropical storms worldwide
Displays current warnings & images
(http://members.tripod.com/~Post_119_Gulfport_MS/tropical.html)

Landslides

The International Landslide Research Group (ILRG)
Information on landslide research
(http://ilrg.gndci.pg.cnr.it/)

Recent developments in landslide mitigation techniques
(http://www.geolith.com/publications/recent_devel/recent_devel.htm)

The U.S. Geological Survey (USGS)
(http://landslides.usgs.gov/html_files/landslides/usgsnoaa/index.html)

Maps/datasets/observing systems

World Conservation Monitoring Centre (WCMC)
Provision of relevant environmental information during emergencies, reports on current incidents
(http://www.wcmc.org.uk/reference/copyright.html)

Natural hazards data
NOAA National Data Centers
(http://www.ngdc.noaa.gov/seg/hazard)

National Centers for Environmental Prediction (NCEP)
Marine, storm, tropical, weather predictions, modelling (http://www.ncep.noaa.gov)

Disaster warning network
Early warnings for earthquakes, tornadoes, lightning storms, tsunami, floods, wild-fires, and all other natural disasters
(http://www.disasterwarning.com)

Geographical Information System, GIS

The GIS Portal

A lot on GIS (http://www.gisportal.com)

ESRI/FEMA (Federal Emergency Management Agency)

Joint site to provide multi-hazards maps and information (http://www.esri.com/hazards)

ESRI

GIS and mapping software (http://www.esri.com)

Satellite

Earth observation for identification of natural disasters

(http://www.kayser-threde.de/ceo/exec.htm)

Real Time Satellite Data Animations (RAMSDIS) online

(http://www.cira.colostate.edu/RAMM/Rmsdsol/main.html)

Appendix 5
Emergency response—international organizations

Joint UNEP/OCHA Environmental Unit
Provides practical assistance to countries affected by environmental disasters (http://www.reliefweb.int/ocha_ol/programs/response/unep/)

United Nations Development Programme (UNDP)
Emergency Response Division—a major partner (http://www.undp.org/erd)

International Strategy for Disaster Reduction (ISDR)
General and specialized information on disaster reduction (http://www.unisdr.org)

IAEA—and its emergency response system
(http://www.iaea.org/worldatom/inforesource/factsheets/emergency)

WHO's division on emergency and humanitarian action homepage
(http://www.who.int/eha/)

World Disasters Report
The International Federation of Red Cross and Red Crescent Societies
(http://www.ifrc.org/pubs/wdr/)

World Bank Disaster Management Facility
Operational support, capacity building, and partnerships with the international and scientific community working on disaster issues.
(http://www.worldbank.org/html/fpd/urban/dis_man/dis_man.htm)

PAHO
Disaster Humanitarian Assistance
(http://www.paho.org)

Habitat
The Habitat Agenda on Disaster prevention, mitigation and preparedness, and post-disaster rehabilitation capabilities
(http://www.unhabitat.org/agenda/ch-4c11.html)

INCEDE
International Center for Disaster-Mitigation Engineering, University of Tokyo
(http://incede.iis.u-tokyo.ac.jp/index.html)

EPC
Emergency Preparedness Canada
(http://www.epc-pcc.gc.ca)

Asian Disaster Preparedness Center (ADPC)
(http://www.adpc.ait.ac.th)

Training/Education

UNITAR
Training and capacity building programmes in chemicals and waste management
(http://www.unitar.org/cwm)

Emergency Management Guide for Business and Industry
(http://www.fema.gov/library)

Disaster Preparedness and Response Bureau (DPRB)
Increase the disaster response capability of fire departments and first response groups through courses, technical assistance, exchange of information (http://www.metro-dade.com/firerescue/disaster.htm)

Appendix 6
References used in preparing this Handbook

Lessons Learned from the Marcopper Tailing Spill. John Loney and Christopher Sheldon, Placer Dome Group. North American Mining, August/September 1998, pp16–20.

Proactive Crisis Communication, David Guth, June/July 1995 (http://www.iabc.com/cw/guth.htm).

Storage of Hazardous Materials: A technical guide for safe warehousing of hazardous materials, Technical Report No. 3. UNEP IE. ISBN 92-807-1238-1.

Cyanide Spill at Baia Mare, Romania. UNEP/OCHA Assessment Mission, March 2000 (http://www.unep.ch/roe/baiamare.htm).

Hazard Identification, Hazard Classification and Risk Assessment for Metals and Metal Compounds in the Aquatic Environment. Peter Chapman, ICME, 1996 (http://www.icme.com).

Risk Assessment and Risk Management of Non-ferrous Metals—Realizing the Risks and Managing the Benefits. ICME, 1997.

International Workshop on Risk Assessment of Metals and their Inorganic Compounds. ICME, 1996.

A Guide to Risk Assessment and Risk Management for Environmental Protection. DoE HMSO, 1995. ISBN 0-11-753091-3.

Trail Community Lead Task Force (Canada): A Co-operative Approach to Community Risk Management. Steven R. Hilts and Terry L. Oke.

Audit and Reduction Manual for Industrial Emissions and Wastes. Technical Report Series No. 7, UNEP and UNIDO, 1991. ISBN 92-807-1303-5.

Environmental Aspects of Selected Non-ferrous Metals (Cu, Ni, Pb, Zn, Au) Ore Mining A Technical Guide. Technical Report Series No. 5. UNEP/IEPAC and ILO, 1991. ISBN 92-807-1366-3.

Health Aspects of Chemical Accidents. Guidance on Chemical Accident Awareness, Preparedness and Response for Health Professionals and Emergency Responders. OECD Environment Monograph No. 18. UNEP IE/PAC, Technical Report No. 19, Paris, 1994.

Hazard Identification and Evaluation in a Local Community. Technical Report No. 12. UNEP IE, 1992. ISBN 92-807-1331-0.

Management of Industrial Accident Prevention and Preparedness: A Training Resource Package. UNEP IE, 1996. ISBN 92-807-1609-3.

Proceedings of the Workshop on Risk Management and Contingency Planning in the Management of Mine Tailings, Buenos Aires, Argentina. ICME and UNEP, November 5 and 6, 1998. ISBN 1-895720-30-3

Report of the OECD Workshop on Risk Assessment and Risk Communication in the Context of Chemical Accident Prevention, Preparedness and Response. OECD, 1997.

TransAPELL: Guidance for Dangerous Goods Transport Emergency Planning In a Local Community. Technical Report No. 35. UNEP DTIE. ISBN 92-807-1907-6.

IMO/UNEP Consultation Version—APELL for Port Areas: Preparedness and Response to Chemical Accidents in Ports, 1996.

APELL: Awareness and Preparedness for Emergencies at Local Level, A Process for Responding to Technological Accidents. UNEP, 1988. ISBN 92-807-1183-0.

APELL Worldwide. UNEP, 1995. ISBN 92-807-1527-5.

A Guide to the Management of Tailings Facilities. The Mining Association of Canada, November 5 and 6, 1998.

Investigation into the Mercury Spill of June 2, 2000 In the Vicinity of San Juan, Choropampa and Magdalena, Peru. Compliance Advisor Ombudsman, July 2000.

A Guide to Tailings Dams and Impoundments: Design, Construction, Use and Rehabilitation. Bulletin 106, UNEP and ICOLD, 1996.

Risk Assessment and Risk Management of Non-Ferrous Metals: Realizing the Benefits and Managing the Risks. ICME, 1997. ISBN 1-895720-19-2

Hazardous Materials Management, Storage and Disposal. Best Practice Environmental Management in Mining Series. Environment Australia, 1997. ISBN 0-642-19448-3 in series 0-642-19418-1.

Water Management. Best Practice Environmental Management in Mining Series. Environment Australia, 1996.
ISBN 0-642-546231 in series 0-642-19418-1.

Cyanide Management. Best Practice Environmental Management in Mining Series. Environment Australia, 1998.
ISBN 0-642-54563-4 in series 0-642-19418-1.

Environmental Risk Management. Best Practice Environmental Management in Mining Series. Environment Australia, 1999.
ISBN 0-642-546304 in series 0-642-19418-1.

Emergency Preparedness and Response in the Mining Industry. Information Note. UNEP/ICME Workshop. Brussels, 29 May 2000.

Australian Minerals Industry Code For Environmental Management. Minerals Council of Australia, February 2000.
(www.enviro-code.minerals.org.au)

Developing and Piloting New Stakeholder Models: The Community and Business Forum in Kyrgyzstan. Mehrdad Nazari, Aug 2000. Submitted to LEAD Cohort 7 Globalisation and Sustainability: Impacts on Local Communities, 13–14 Aug 2000, Vancouver, Canada.

Lima Workshop on Mining and Sustainable Development in the Americas, June 27–29, 1998 Peru. Report of Proceedings. International Institute for Sustainable Development and International Development Research Centre.

Tailings Dams, Risk of Dangerous Occurrences. Lessons Learnt from Practical Experiences. ICOLD/UNEP Bulletin 2001.

Crisis Communications: Guiding Principles. European Chemical Industry Council Book of Best Practice. Marc Devisscher, Oct 1993.

Responsible Care Guidelines—Crisis Communications Guidelines of the European Chemical Industry Council. (http://www.cefic.be)

Risk Management and the Future. Ed.Tom Beer. Australian Minerals and Energy Environment Foundation 2000.

Mudder, T.I & Botz, M.M A Global Perspective of Cyanide. Published in Workshop on Industry Codes of Practice: Cyanide Management Report. UNEP DTIE and ICME, May 2000.

About the UNEP Division of Technology, Industry and Economics

The mission of the UNEP Division of Technology, Industry and Economics (UNEP DTIE), is to help decision-makers in government, local authorities, and industry develop and adopt policies and practices that:

- are cleaner and safer;
- make efficient use of natural resources;
- ensure adequate management of chemicals;
- incorporate environmental costs;
- reduce pollution and risks for humans and the environment.

The UNEP DTIE, with its head office in Paris, is composed of one centre and four units:

- **The International Environmental Technology Centre (Osaka)**, which promotes the adoption and use of environmentally sound technologies with a focus on the environmental management of cities and freshwater basins, in developing countries and countries in transition.
- **Production and Consumption (Paris)**, which fosters the development of cleaner and safer production and consumption patterns that lead to increased efficiency in the use of natural resources and reductions in pollution.
- **Chemicals (Geneva)**, which promotes sustainable development by catalysing global actions and building national capacities for the sound management of chemicals and the improvement of chemical safety world-wide, with a priority on Persistent Organic Pollutants (POPs) and Prior Informed Consent (PIC, jointly with FAO).
- **Energy and OzonAction (Paris)**, which supports the phase-out of ozone depleting substances in developing countries and countries with economies in transition, and promotes good management practices and use of energy, with a focus on atmospheric impacts. The UNEP/RISØ Collaborating Centre on Energy and Environment supports the work of the Unit.
- **Economics and Trade (Geneva)**, which promotes the use and application of assessment and incentive tools for environmental policy and helps improve the understanding of linkages between trade and environment and the role of financial institutions in promoting sustainable development.

UNEP DTIE activities focus on raising awareness, improving the transfer of information, building capacity, fostering technology cooperation, partnerships and transfer, improving understanding of environmental impacts of trade issues, promoting integration of environmental considerations into economic policies, and catalysing global chemical safety.

UNEP DTIE operates **Mineral Resources Forum—Environment** website.

http://www.mineralresourcesforum.org

Mineral Resources Forum—Environment is an Internet framework for environmental perspectives of information on the theme of minerals, metals and sustainable development, promoting the exchange of knowledge, experience and expertise in the impact of mining, mineral processing and metals on the natural environment, bringing together governmental and intergovernmental actors, resource companies and other concerned organisations and persons from civil society.

For more information please contact:
United Nations Environment Programme
Division of Technology, Industry and Economics
Tour Mirabeau 39–43 quai André Citröen
75739 Paris Cedex 15, France
Tel: +33 1 44 37 14 50,
Fax: +33 1 44 37 14 74
E-mail: unep.tie@unep.fr
http://www.uneptie.org

LOUIS - JEAN
avenue d'Embrun, 05003 GAP cedex
Tél. : 04.92.53.17.00
Dépôt légal : 325 – Mai 2001
Imprimé en France